THE PACIFIC RIM

by Hyung Woong Pak
with Dorothy and Thomas Hoobler

Consultants

A. Elgin Heinz
Consultant, Education about Asia
San Rafael, California

N. T. Wang
Professor of Business
East Asia Institute
Columbia University

SCHOLASTIC INC.

Allen County Public Library
Ft. Wayne, Indiana

Titles in This Series
CANADA
CHINA
GREAT BRITAIN
THE INDIAN SUBCONTINENT
JAPAN
LATIN AMERICA
MEXICO
THE MIDDLE EAST
THE PACIFIC RIM
SOUTHEAST ASIA
THE SOVIET UNION AND EASTERN EUROPE
TROPICAL AND SOUTHERN AFRICA
WESTERN EUROPE

No part of this publication may be reproduced in whole or in part, or stored in a retrieval system, or transmitted in any form or by any means, electronic, mechanical, photocopying, recording, or otherwise, without written permission of the publisher. For information regarding permission, write to Scholastic Text Division, 730 Broadway, New York, NY 10003.

ISBN 0-590-34927-9

Copyright © 1990 by Scholastic Inc.
All rights reserved.
Published by Scholastic Inc.
Printed in the U.S.A.

12 11 10 9 8 7 6 5 4 3 2   0 1 2 3/9

Hyung Woong Pak is a native of Korea and has a bachelor's degree in economics from the University of Chicago. There he did graduate work with the Committee on Social Thought. He edited *Chicago Review* for many years. He was also president of a publishing company in New York. At present, Pak is a member of the Asian Task Force of Philadelphia School District and also a member of the Board of School Directors, Cheltenham Township, Pennsylvania.

Dorothy and Thomas Hoobler have written numerous books on current affairs and history, including books on Japan, China, and U.S.–China relations. They have also written many biographies, including one on Zhou Enlai.

Publisher: Eleanor Angeles
Editorial Director: Carolyn Jackson
Project Editor: Richard G. Gallin
Art: Harry Chester Associates
Photo Research: Rosalyn Sohnen
Maps: David Lindroth
Contributing Writers: Lisa Keating, Susan Merrill, Penny Parsekian

---

COVER: "Boats of every description—huge cargo ships, hydrofoils, ferries, and ramshackle fishing vessels ply one of the world's great harbors. On land, Hong Kong's distinctive skyline soars with sleek skyscrapers that house banks and insurance and trading companies. They form the basis of Hong Kong's great financial empire." —Chapter 9

THE
PACIFIC RIM

Table of Contents

Prologue: Traveling into the Future   9

PART 1: TRADITION AND CHANGE   16

Chapter 1: The Land   17
Chapter 2: A Mosaic of People   33
More About Confucius   35
More About Buddhism   44
Chapter 3: History   53

PART 2: THE BIG DRAGON   66

Chapter 4: Japan: The Making of an Economic Miracle   67
More About Shinto   78
Chapter 5: From Military Disaster to Economic Triumph   85
Chapter 6: Living in an Economic Miracle   103

PART 3: THE FOUR TIGERS   122

Chapter 7: South Korea: The Miracle on the Han   123
More About South Korean Junior and Senior High School   138
Chapter 8: Taiwan: We Will Make Anything You Want   143
More About Sports in Taiwan: Shuttlecock Kicking and Baseball   158
Chapter 9: Hong Kong: Living on Borrowed Time   163
Chapter 10: Singapore: A Mighty Ministate   175
More About Asian Names   183

PART 4: ON THE SHORES OF THE PACIFIC	186
Chapter 11: The Little Tigers: Natural Wealth, Growing Industries	187
More About Islam	198
Chapter 12: Australia and New Zealand: Diggers and Kiwis	203
Chapter 13: Developing Countries of the Pacific Rim	217
Chapter 14: The Eastern Pacific Rim	231
Epilogue: The Pacific Century	245
Pronunciation Guide	247
Index	250
Photo Credits	256

Maps:

The Pacific Basin	6
Countries of the Western Pacific Rim	18
Geography and Agriculture of the Pacific Rim	25
Major Natural Resources and Industries of the Pacific Rim	30
Population Distribution in the Western Pacific Rim	42
Colonial Powers in the Pacific Rim, 1914	56
The Height of the Japanese Expansion	100
The Economy of Japan	108
The Korean Peninsula	124
The Little Tigers	199
Australia and New Zealand	215
Average Income per Person in Pacific Rim Economies	227

THE PACIFIC BASIN

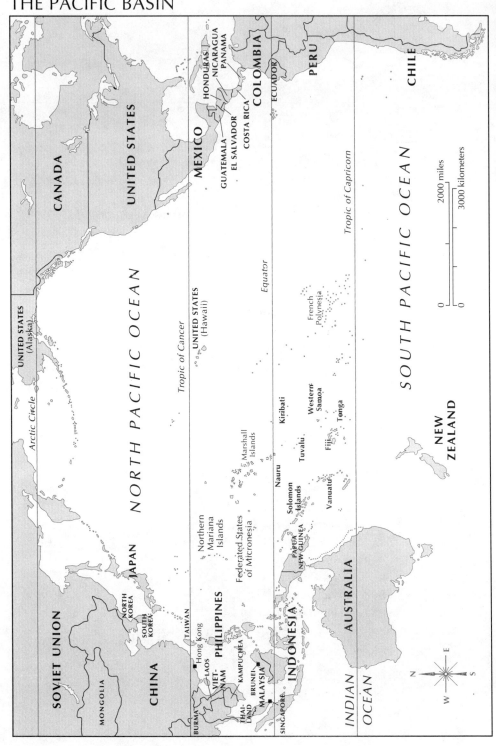

There is rhythm in everything. . . . The rhythms of the path of dance. . . . and of the wind and string instruments are among the commonly known and obvious ones. . . .

. . . In the path of commerce, too, there is the rhythm by which one becomes wealthy, and the rhythm by which the wealthy go bankrupt, with the differences in the rhythms according to each path. The rhythm with which things progress and the rhythm with which things deteriorate should be understood and their differences recognized.

MIYAMOTO MUSASHI,
Japanese artist-soldier, 1645

PROLOGUE

TRAVELING INTO THE FUTURE

IF YOU GOT INTO AN AIRPLANE and flew across the Pacific Ocean to Asia, you would land at your destination a day ahead of United States time. You would be moving toward tomorrow throughout your flight. Many people believe that the Pacific region is where the future lies and that the world is moving toward "The Pacific Century."

Why is there so much interest in the Asian countries that are on the western edge of the Pacific Ocean? They are the heroes of a startling economic story. From the 1970s, the wealth in Asian economies such as Japan, South Korea, Taiwan, Singapore, and Hong Kong has

An electronics factory in Hong Kong. By the mid-1980s, the combined microelectronics exports of Japan, Malaysia, Singapore, and South Korea were greater than those of the United States.

grown faster than in any other region of the world. These so-called Pacific Rim countries are the engine of the world economy today. Factories in South Korea and Taiwan are turning out cars, computers, and video cassette recorders; the booming ports of Singapore and Hong Kong (a British colony) send ships carrying toys, fashions, and electronic products to the rest of the world. Some experts believe that by the year 2000, the nations around the Pacific—both eastern (the United States and Canada and parts of Latin America) and western (the Pacific Rim in Asia)—will produce more than 55 percent of all the goods and services in the world.

This startling development has brought a rapid change in people's ideas about Asian goods. In the past, products stamped "Made in Japan" or "Made in Taiwan" were regarded as shoddy. Today Asian trademarks such as Sony, Toyota, and Hyundai represent quality to consumers all over the world. Moreover, companies that used to be thought of as "all-American," like General Electric, IBM, and Chrysler, now sell many products that have been made in Asia.

The rise of the Pacific Rim has in some ways moved the center of the world. Until recently, the nations around the Atlantic Ocean—the United States and countries in Western Europe—were the political and economic leaders of the world. Maps of the world from the 19th century placed Europe at the center. This is the reason why we use such terms as Far East for the Asian countries of the Pacific Rim. They were east of Europe. But today, Atlantic dominance has declined. By 1978, the United States had greater trade with Asia than it had with Europe. Economically, the United States is now more a Pacific nation than an Atlantic one. That trend will continue.

This book is an introduction to the Pacific Rim. It will concentrate on the Asian, or western side, of the Pacific.

The main trading board of the Tokyo Stock Exchange. What happens to prices on the Tokyo Stock Exchange now closely affects what happens to prices on the New York Stock Exchange—and vice versa.

A busy street in Tokyo, considered the world's largest city. The capital of Japan has over 80,000 streets. Do you think it would be easy to be a mail deliverer in Tokyo?

The nations there blend ancient tradition with modern economies. We will seek to find the reasons for their phenomenal success.

Miracle economies. The capital of the western Pacific Rim is Tokyo.* For it is Japan that has led the way. From the ashes of their defeat in World War II, the Japanese people have built the most dynamic economy in the world. From the bombed rubble of Tokyo rose the skyscrapers of today. Within those glass and steel towers are the headquarters of more than half a million corporations. Japanese products are the envy of the world. Japan is the first of the miracle economies.

Other Asian economies have been inspired by Japan's success. South Korea, Taiwan, Hong Kong, and Singapore—known as the Four Tigers or Little Dragons—have followed Japan's lead. (Japan is the Big Dragon.) Just as Japan built its economic success with workers who were paid less than Americans and Europeans, the Four Tigers have taken advantage of labor that is lower-paid than Japan's. They have entered the international economy in a big way, producing their own miracle economies. Today they are competing with Japan in producing automobiles, steel, electronics, and many other products.

A sign in Taipei,* the capital of Taiwan, sums up the attitude of the Four Tigers: "We Will Make Anything You Want to Buy." And they do—either better or cheaper or both. They are shining examples of newly industrialized economies (or NIEs).

The success of the Four Tigers has fueled development in the so-called Little Tigers—Thailand,* Malaysia,* Indonesia, and the Philippines. These countries are beginning to develop their own industries to

---

*See Pronunciation Guide, which begins on page 247.

make export goods. Rounding out the Pacific picture are the countries of Australia and New Zealand, whose economies have also benefited from the prosperity of their neighbors.

Cooperation or competition. Americans are the biggest customers of the Pacific Rim. Televisions, video cassette recorders, running shoes, and automobiles, which all used to be made in the United States, are now made in the western Pacific Rim. The parts inside many American-made products are also imported from the western Pacific.

The success of the Pacific Rim has rested on making products cheaper and more efficiently than they can be made in the United States. Some workers in American factories who used to turn out televisions, stereos, and automobiles have lost their jobs. In industries such as consumer electronics products, American manufacturers have given up trying to compete. Virtually all the products they sell are imported from the other side of the Pacific.

Though Americans are the best customers for Asian-made products, Asians buy few American-made goods. For several reasons, it has been difficult for American companies to sell their products in the Pacific Rim countries. The United States has consistently bought more from these countries than it has been able to sell to them, causing a trade deficit. The United States used to export more goods than it imported. Now the situation is reversed.

Most Pacific Rim nations traditionally have been friends and strategic allies of the United States. Indeed, over the past 40 years, such nations as South Korea, Japan, Taiwan, and the Philippines have relied on the United States' armed forces to defend them. The United

South Korean autos ready to be shipped abroad. In 1980, South Korea produced only a few thousand cars. Today, only eight other countries make more autos than South Korea does.

States encouraged free trade with these countries, and even helped them to develop industries.

Past experience has shown that free trade between nations encourages a stable world economy. But the economic success of the Pacific Rim countries has caused the United States to reconsider its relationship with them. How long can it afford the growing trade deficits? Should it put a tariff (tax) on foreign goods to raise their prices and in this way restrict imports? The answers to these questions are not easy to find—but they will be important in shaping the world of the 21st century.

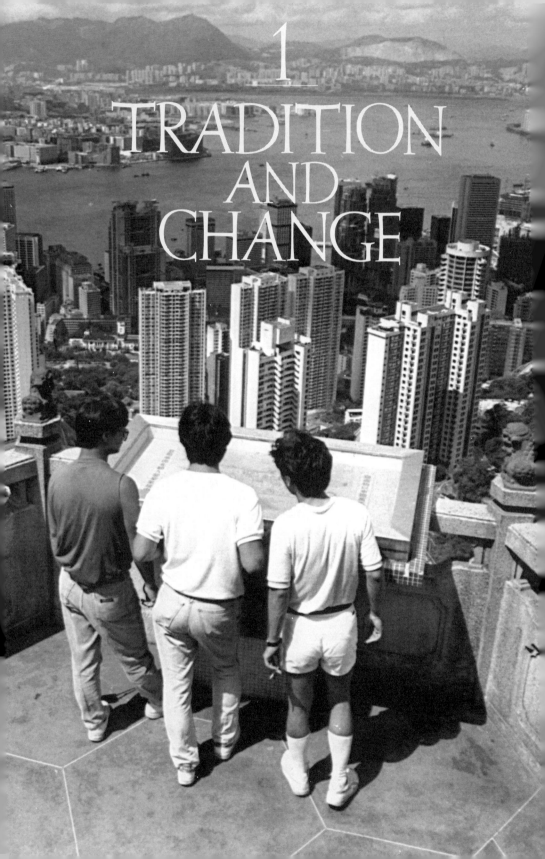

1
TRADITION AND CHANGE

Chapter 1

The Land

THE COUNTRIES OF THE PACIFIC RIM lie on either side of the world's largest ocean—the Pacific. Occupying nearly half the surface of the earth, the Pacific is dotted by only a few tiny islands. Tokyo, Japan, is almost directly west of Los Angeles, but the two cities are separated by 5,470 miles of water—and that is not the Pacific's widest point.

A new way of looking at the world. Most maps of the world have had Europe at their center. Asia was called the Far East because on such maps Asia was at the right edge, the easternmost edge of the map. The Americas were placed on the left side of the map, on the westernmost edge. Such maps were most common through the early part of this century when the British Empire was at its height. Britain controlled colonies that

A view of downtown Hong Kong and Victoria Harbor from Victoria Peak. Hong Kong is one of the most important ports in the western Pacific Rim region.

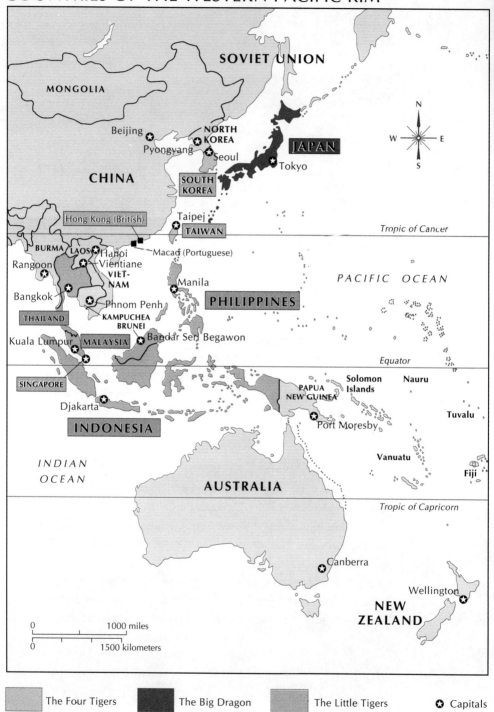

covered almost one-fourth of the globe's land. Just as East Asia and Southeast Asia were called the Far East by Europeans, the region of southwestern Asia was called the Near East or the Middle East

Now look at the map on page 6. This gives a picture of the earth from a different viewpoint. It shows the earth from a viewpoint centered on the mid-Pacific Ocean. And the map on page 18 shows a close-up of the western Pacific Rim, the countries that are on the western edge of the Pacific Ocean. That is the way we will be looking at the region. It is an important part of the story of the Pacific Rim. You will be able to picture more easily the relationships among parts of the Pacific Rim by thinking of the region in this new way.

Many of the countries on the Asian Pacific Rim are islands and peninsulas. Indonesia is the largest archipelago,* or island group, in the world—with over 13,500 islands. The Philippines consist of over half that many islands. Japan, Taiwan, and New Zealand are also island nations. The three nations on peninsulas are Korea, Thailand, and Malaysia. (But part of Malaysia is also on a large island.)

The Pacific Rim countries present a crazy quilt assortment. They vary greatly in size. Australia is the only nation in the world that is also a continent. It is almost as large as the United States. Singapore is a city-state smaller in area than New York City. South Korea is the southern half of what was once Korea. Some, indeed, are not even independent nations. Hong Kong is a colony of Great Britain. Taiwan itself is regarded as part of China. The People's Republic of China overshadows all of them in land mass and population, but it is a "sleeping giant"— so far, not as important a player in the world economy.

Almost every type of landform is found within the

Yeh-Liu (Wild Willows) Beach on the northern coast of Taiwan. Visitors come to see the strangely shaped rocks carved by wind and water. What other Pacific Rim countries are island nations?

Pacific Rim. Australia's huge interior is desert. Three-fourths of Indonesia is covered by vast rain forests, which are also found in Malaysia, Thailand, and the Philippines. Both Japan and the Philippines were formed by lava forced up from undersea volcanoes. Today, Japan, Indonesia, and the Philippines still have active volcanoes. Some areas also experience earthquakes as the shifting plates move underneath the Pacific.

Mountains dominate many regions. Japan's Mount Fuji is a symbol of the country's strength. New Zealand's South Island has its Southern Alps—many mountains over 10,000 feet high. Several of Indonesia's larger islands are mountains that rise out of the sea. The part of Malaysia on a peninsula is divided into east and west by a mountain chain running through its center.

Though four-fifths of Japan's land area is too rocky for farming, other nations have an abundance of rich soil. In Thailand, the Chao Phraya River* flows south to the sea, carrying fertile silt that produces some of the richest rice paddies in the world. Even in Singapore, the city-state, people still use precious land for growing rice, rubber trees, tropical fruits, and vegetables.

Snow-capped Fuji-no-Yama (Mount Fuji) looms in the distance, miles away from these apartment buildings and factories. In summer, many Japanese make the six- to eight-hour nighttime climb in order to watch the sunrise from its peak.

Climate. The most important weather report in Asia is the prediction for the arrival of the monsoon.* The monsoons are seasonal winds that blow from the Asian mainland part of the year and from the ocean in another part of the year. When these strong winds blow across water, they pick up moisture that falls as rain or snow. In the summer, the mainland of Asia heats up, and the wind and rain rush in from the cooler sea. In the winter, the land cools down and sends the wind back over the warmer sea.

From Japan to northern Australia, the coming of the monsoon changes daily life. In Bangkok, Thailand, the streets are filled with water during the monsoon. People customarily leave their shoes at home during the rainy season, and shopkeepers build small dams in their doorways. Rice farmers welcome the monsoon rains, but when the rain is too heavy, serious flooding can result. Extremely heavy monsoon rains can force fishing boats to remain on shore. Even large ocean-going shipping vessels are affected, and during the monsoon season Singapore may have to adjust the shipping schedules in its busy harbor.

In many parts of Indonesia, it rains just about every day. Because the nation straddles the Equator, there are no seasons. Throughout the year, the temperature and humidity remain high. In contrast, Japan and South Korea have four distinct seasons like the eastern United States.

Ocean currents also affect the climate of several western Pacific Rim countries. Some ocean currents carry warm water into parts of the ocean that are colder. The Japan Current, or Kuroshio,* is such a current. It flows northward along Japan's southeastern coast, making the temperature mild. A cold current, called the Oyashio* Current, flows southward and cools the northeastern shores of Japan. Australia's northern and eastern

22

Floods slow traffic on a main street in Bangkok, Thailand, where the monsoon season starts in May or June and lasts about five months. During these months the weather alternates between sudden rainstorms and sunny skies.

coasts are warmed by currents. Cold currents from near Antarctica cool its southern coast.

Because of their location south of the Equator, the seasons in Australia and New Zealand are reversed from what they are in the Northern Hemisphere. In July and August, when lands north of the Equator are experiencing summer, their people shiver in winter's snow and ice. Southern Australia and New Zealand's South Island are colder than the north. Monsoon rains produce jungle growth in northern Australia. Southern New Zealand has glaciers,* huge mountains of ice that form on land and fall into the sea.

Western Pacific Rim farmers make use of fertile land by planting their crops on terraced hillsides.

Country and city. For thousands of years, rice farmers have provided the major food for the people of East Asia. In the Asian countries of the Pacific Rim, land suited for rice farming is scarce and precious. In many countries, farmers shape the hills into layers of flat terraces to use every possible inch of land.

Rice growing requires intense human labor. Rice must be grown in shallow water, and dams and channels must be built to control the water supply. Rice seedlings are planted in groups, and when they reach a proper size, taken up and replanted farther apart. Weeds must be pulled out by hand. No machinery is yet available in most places that can do these things in several inches of water—so the villagers spend hours of backbreaking labor tending the rice paddies daily.

However, improved types of rice have increased the

yields. In some countries, such as Thailand and Indonesia, two or three crops a year are grown and harvested. Thailand's farmers are so successful that the country has become the fifth largest producer of rice and one of the world's largest food exporters.

Because of the long coastlines of the island and peninsula nations, ocean fishing is also an important source of food. A favorite dish in many nations is a bowl of rice covered with a sauce made from salt and fish. Soybeans are also an important food. Farmers in Indonesia, North and South Korea, Thailand, Japan, and Vietnam grow large amounts of this protein-rich food.

Brazilian rubber trees were brought to the western Pacific Rim in the 1870s. Workers in Indonesia and Malaysia tap the latex (natural rubber) by making a special cut in the tree trunk and then waiting about four hours before collecting the latex.

Australia's climate and large area make it possible to grow wheat, which is an important export. On large ranches, Australia's cowboys also tend cattle and sheep. The green hills of New Zealand provide good grazing for its sheep industry. Wool is an important export, and shoppers in Europe and the United States can enjoy eating New Zealand spring lamb during the Northern Hemisphere's fall and winter.

Some Pacific Rim nations produce other agricultural products such as rubber, sugar, palm oil, tea, and coffee. Most of these are grown on large plantations in Southeast Asian countries, especially Malaysia and Indonesia. One of the most important cash crops is the opium poppy, grown in northern Thailand and adjoining areas of Laos and Burma. Most of this crop is exported and used to make dangerous, illegal drugs sold at high prices around the world.

Throughout Asia, the village has traditionally been the focus of life. Most of the people have been farmers who lived in rural villages. But as the Pacific Rim nations have developed industries, a massive shift to the cities has taken place.

The rate of urbanization* varies—less than one-fifth of Thailand's population lives in cities, whereas Singapore is almost entirely an urban nation. Seoul, the capital of South Korea, has grown so quickly that it is now the world's fifth-largest city. About one out of every five South Koreans lives there.

Urban areas in many of the Pacific Rim nations have grown in a similar fashion. The economies of the Pacific Rim include some of the most exciting cities of the world. The skyscrapers by Hong Kong's magnificent harbor are a thrilling sight. People can view Sydney, Australia's great port city and harbor, from a bridge with the longest concrete arch in the world. The dense office buildings of downtown Singapore reflect the fact that it exports more manufactured goods than all of China, the world's most populous nation. Bangkok's night life makes it a tourist attraction for people from other Asian nations.

Natural resources. In western Europe and North America, powerful industrial countries developed because they had coal and iron that could be used to produce steel. Energy to run factories came at first from the flowing rivers, and then from coal, oil, natural gas, and hydroelectric power (electricity produced by the flow of water). Nations that did not have these natural resources had to find ways to get them if they wanted to develop. Some methods were trading, waging war, or obtaining colonies.

One of the most amazing things about the industrial development of the Asian countries of the Pacific Rim is that they have almost none of these natural resources. Of the Four Tigers, only South Korea has an important natural resource—coal deposits. The Four Tigers' secret weapon—their most important resource—has been an industrious and highly educated people.

However, the Little Tigers have more traditional kinds of natural resources. Indonesia, Malaysia, and Thailand are world leaders in rubber production. These three countries also have rich deposits of tin. Indonesia has still-untapped sources of iron, coal, copper, and uranium, and is a major producer of oil. Though they have not yet caught up with the Four Tigers, these developing nations have the resources to surpass them. Australia, because of its size, has many resources needed in modern industry: coal and minerals such as iron, tungsten, tin, manganese, nickel, and bauxite.* It, too, has a brilliant future waiting to happen.

Double-check

Review

1. What is the meaning of the term *Pacific Rim*?
2. What important Pacific Rim countries are located on peninsulas?
3. Name the Four Tigers of Asia.
4. Why did Europeans call East Asia the Far East?
5. What are five of the major natural resources of the Pacific Rim?

Discussion

1. This chapter points out that where people live influences how they live. List the ways this is true for your community. Then compare your list with the information about the Pacific Rim in this chapter. Would it make sense to say, "Geography is destiny"? Why or why not?

2. Almost all the Asian nations of the Pacific Rim have at least one border on an ocean or sea. In what ways does access to water affect life in the Pacific Rim? What problems might there be for a landlocked country such as Laos?

3. List several of the natural resources located in your state or province. Then list how and by whom, and for what purpose these resources are used. Compare your list with a list of the natural resources discussed in this chapter. Make a prediction about how you think the countries of the Pacific Rim use their natural resources. Keep a record of these predictions and compare them with the information you gain by reading the other chapters.

Activities

1. A committee of students might be formed to prepare a large wall map of the Pacific Rim for use with this and future chapters. They could use the map on page 11 as a guide, and then add information to it from other maps, including others in this book.

2. Eight words in Chapter 1 are starred (*). This indicates that they are in the Spelling and Pronunciation guide at the back of the book. A committee might assume primary responsibility for teaching other students how to pronounce and define these words. They could do this, in advance, for all chapters.

3. Form committees of four or five students each to do research on natural resources. Each group should select a different natural resource mentioned in this chapter and do library research and write a report about the ways that the particular resource is used in modern industrial countries. Each group should present its report to the class. The class should vote on which three resources they think are most important for a modern society.

MAJOR NATURAL RESOURCES AND INDUSTRIES OF THE PACIFIC RIM

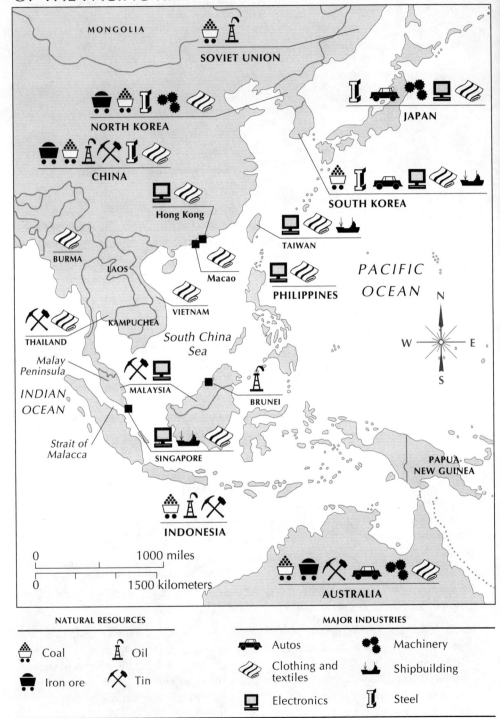

Skills

READING A MAP

Study the map on page 30. Then answer the following questions.

1. On what peninsula are major tin resources located?
   a. Malay Peninsula   b. Kamchatka Peninsula   c. Alaskan Peninsula

2. What major archipelago nation is an important source of oil?
   a. Japan   b. Indonesia   c. Philippines

3. Ships bringing oil to Taiwan from lands in southwestern Asia probably have to sail through what body of water to get from the Indian Ocean to the South China Sea?
   a. Pacific Ocean   b. Atlantic Ocean   c. Strait of Malacca

4. Which countries have large iron ore deposits?

5. Suppose Japan needs coal to produce steel. What countries in this part of the Pacific Rim might sell it coal?

Chapter 2

A Mosaic of People

MEET VICTOR CHUA. HE IS a 17-year-old student in Singapore, the smallest of the Pacific Rim nations. Victor goes to National Junior College, one of Singapore's toughest high schools. You wouldn't guess how bright its students are by looking at them—most of them seem sleepy. Victor explains: "Most of us have been up all night studying. Exams are only 14 weeks away."

Why does Victor study so hard? It's not just because he wants to get a good job. He agrees with his nation's leader, Prime Minister Lee Kuan Yew, that Singapore's people must be the best and brightest in Asia. Victor says, "Singapore has no oil or other resources to sell—just the minds and talents of our people. If we're to compete in the world economy, we must keep learning, training, and bettering ourselves."

High school students in Taiwan. The peoples of the western Pacific Rim are of many different ethnic, cultural, language, and religious groups.

That's the reason for the recent success of Japan and the Four Tigers—their people. Like three-fourths of the people of Singapore, Victor is of Chinese descent. Taiwan is entirely Chinese, and Hong Kong nearly so. Though the people of Japan and Korea are not Chinese, their cultures were strongly influenced by China. And one of the strongest forces in Chinese culture is Confucianism.

The Confucian ethic. Confucius (Kongfuzi in pinyin spelling) was a Chinese philosopher who lived around 500 B.C. The ancient Chinese turned his ideas into a way of life. His philosophy spread to Korea and Japan. The Chinese who moved throughout Southeast Asia carried Confucius's ideas with them.

This way of life, called the Confucian ethic, is behind the success of today's Pacific Rim. It is based on three important values. The first one is simple: work hard. In Japan and the Four Tiger nations, people work longer hours than most Americans do. If the job is not finished, people stay overtime without complaining. It is rare for anyone to take a day off because of illness. Employees often willingly give up their vacations if they are needed at work. The school year is longer too—about two months longer than in the United States, including half-days on Saturday.

A recent study showed the difference in attitudes toward hard work and achievement. Students in South Korea, the United States, and other countries were given a mathematics test. The South Koreans scored among the highest. Americans scored among the lowest on the test. Yet when asked how they thought they were doing, most American students thought their math achievement was fine. Only one-fourth of the South Koreans were happy with their test scores. And many Asians believe long hours of study will lead to success.

More About Confucius

Kongfuzi (Confucius) taught that rulers exist to secure the happiness of their subjects and that the family provides the model for all human relations.

"I think I must study harder from now on," young Prince Akihito wrote in his diary. His father, the emperor of Japan, had just announced Japan's unconditional surrender to the Allies, ending World War II.

Today Akihito is emperor. He didn't get there by studying. He became emperor by birth. But the belief that studying leads to understanding, which, in turn, leads to greatness—for nations as well as for individuals—is part of an Asian tradition. The founder of this tradition was a man called Confucius. In pinyin spelling he is called Kongfuzi or Kongzi.

Although he lived and taught about 2,500 years ago, Confucius's ideas still make sense today. This is why he is considered one of the world's great thinkers. Scholars compare him to Jesus and Socrates, yet he didn't found a religion or establish a method of inquiry.

Confucius's main concern was good government. He thought that if humankind was to succeed it would be due to the right kind of leadership. Confucius's ideas about how to govern set high standards that, in fact, few governments have been able to follow. A good leader rules by example, he said, not by punishment. He reasoned that if people were "regulated by penalties" they would merely "try to keep out of jail." Whereas, if people saw excellence in their leaders, they would "have a sense of shame" that would encourage them to try to improve.

To Confucius, the government's most important job was to inspire people. Once Confucius was asked by a disciple what the government's most important responsibilities were. Confucius responded: To feed and protect the people and to gain their admiration. But the most important of these was the last. "For, from old, death has been the lot of all men, but if the people have no confidence in the government, the state cannot stand."

To create the best government, Confucius felt that administrative positions should be "open to the talents." That is, ability, not birth, should determine whether a person gets a government post. Confucius's ideas were finally put into practice about two hundred years after his death. A state examination system was started to select government officials. It survived for over two thousand years.

While good government was the goal of Confucius's teachings, study was the way to achieve it. Confucius believed that humankind had enough experience in governing to draw conclusions about what worked and what didn't. This experience was contained in the court records of the kingdoms of the past. Confucius spent a good part of his life studying and interpreting these documents and urging his followers to do the same.

Relationships among people were of the utmost importance to Confucius. He always seemed to naturally say the right thing. But, in fact, he believed in following formalities of behavior called rituals. He understood how difficult it is to convey the right feelings to others. For example, when a loved one died, he noted that to "treat the dead as dead would show a lack of love and therefore cannot be done; to treat the dead as living would show a lack of wisdom and likewise cannot be done." So certain rituals had been worked out to show respect and love for the dead and the living. Family relationships and friendships, too, both had their special requirements. Confucius knew these rituals by heart.

Maybe Confucius was so admired because he was always striving to improve himself. Everyone had this opportunity, he thought, and most situations give us the possibility to learn and improve. "When walking with a party of three," he once said, "I always have teachers. I can select the good qualities of the one for imitation, and the bad ones of the other for correcting them in myself."

The result of all this study and effort was not to become a stodgy academic or a well-behaved robot. It was personal satisfaction, happiness. "Those who know the truth," he said, "are not up to those who love it; those who love the truth are not up to those who delight in it."

What do we know about Confucius, the man? Fact and fiction are hard to sort out for a man who lived so long ago and left almost no writing. He was born around 551 B.C. in the city of Qufu, China, on the Grand Canal in what is today the province of Shandong. His family had fallen upon hard times, and his father may have died when he was young. From an early age, Confucius had to fend for himself.

Although he worked hard to win a government position, when he finally succeeded, he didn't keep it long. Mostly, he taught. At this he was very successful. Over 3,000 students flocked to his door. By the time he died in 479 B.C., he had disciples who continued his teachings. A disciple several generations later finally wrote down the sayings of the beloved master.

Today, Confucius's birthday is celebrated as Teachers' Day, a national holiday in Taiwan and other Asian countries. It is just one indication of the high regard still reserved for those who teach or study. Close relationships are still defined by formalities. Many of Confucius's ideas have not enjoyed a continuous following. Over the centuries they have been sometimes misunderstood or rejected or misused. But they keep coming back to challenge and inspire people.

One of South Korea's largest businesses provides afterschool schools like this for its workers' children.

The second Confucian value is respect for education. For over a thousand years, the emperors of China picked the brightest people in the country to serve them. Any young man could take the exams that decided who was capable of serving the emperor. But they often had to study for years to pass the exams. Families, and even whole villages, saved their money to pay for the education of a boy who might pass the exams—because he would bring honor to his relatives and village. Passing the imperial exams was like becoming a famous football star in modern America.

Today, though the Chinese emperors are gone, young men and women still bring honor to their families by getting a good education. Students respect their teachers because they help them to learn. Though an Asian elementary-school classroom often has more than 60 students, the teacher has no discipline problems. The students write down everything their teacher says. In Japan, mothers called *kyoiku* mamas,* or "education mamas," attend classes and take notes if their child is sick.

Researchers found great respect for excellent schoolwork when they conducted a recent survey. American and Chinese parents were asked how well their child would have to score if the average score on a test was 70. Most American parents said they would be satisfied

What important Confucian values might be illustrated by the children, parents, and grandparents in this photograph?

if their child's score was a few points above 70. Chinese parents expected more from their children. Chinese students would have to score in the 90s for their parents to be happy.

Family loyalty is the third Confucian value. Children honor their parents, grandparents, and long-departed ancestors for giving them life. Confucius said that it is your responsibility to take good care of yourself and your body, because your parents gave it to you. Teenage crime, unwed mothers, and drug-taking are rare in such countries as Singapore. Any of these things would bring great disgrace on your family.

Respecting elders is not limited to relatives. Even among friends, if the friend is older than you, you usually address him or her as "older brother" or "older sister." Confucius saw respect for others, particularly elders and those in authority, as a way of promoting harmony. These ideals of loyalty and respect are extended to the workplace and the nation. Asia's successful nations are proof that Confucius's ideas still work.

A diversity of religions. The more than 1.7 billion people who live in the Pacific Rim countries are part of many culture groups. They speak hundreds of forms of Chinese, Japanese, Korean, and Malay, as well as English. Millions of people speak Thai, Vietnamese, Tagalog (in the Philippines), Javanese (in Indonesia), and other languages.

The diversity extends to religion as well. Almost every kind of religion is practiced in some parts of the Pacific Rim. Indeed, many people see no conflict in following several religions.

The oldest religious practices of the region are called *animism*. Animism is the belief that gods or spirits make their homes in living things and natural objects. Even such things as brooks, waterfalls, and rocks have their

When the cherry blossoms bloom, thousands of Japanese gather in Tokyo's Ueno Park. Shinto animist beliefs emphasize Japan's traditional love of natural beauty.

spirits. These spirits can influence daily life for good or ill. If a person has what we might call "luck"—good or bad—animists believe it is probably due to the influence of a nearby spirit. Thus, spirit-houses, or shrines to these spirits, still dot the countryside in many countries. People bring flowers or gifts of food to ask the spirits for help. Even the American embassy in Thailand has its own spirit-house.

Buddhism, which began in India, soon came to China, where many people took up its ideas and practices. From

POPULATION DISTRIBUTION IN THE WESTERN PACIFIC RIM

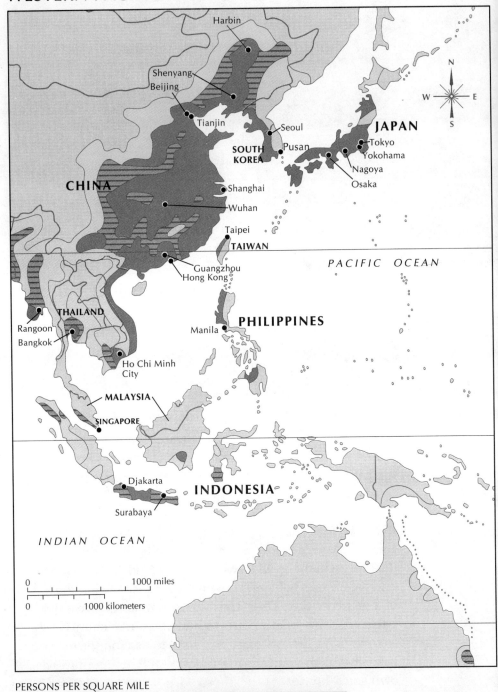

Buddhist monks in Thailand. About 95 percent of Thailand's people are Buddhists.

there, it spread to Korea then Japan, where it developed new forms. Today, Buddhism is stronger in Japan and Southeast Asia than it is in India, the country where it began. Many of the finest examples of Asian art and architecture are Buddhist temples. Some people devote themselves to following a strict Buddhist life. They become monks or nuns, living in monasteries where they study and pray.

Arab sea traders carried the religion of Islam to Southeast Asia. Today, Islam is the dominant religion of Malaysia. And more Muslims live in Indonesia than in any other country in the world. Islam has developed different forms, blending with animism, Buddhism, and the other great Indian religion, Hinduism.

Christianity was a latecomer to the region, carried from Europe in the past 300 years by Christian missionaries. But today it is the majority religion of the Philippines, Australia, and New Zealand. South Korea also has a large community of Christians.

Centuries ago a Chinese emperor invited followers of many different religions to his court. He listened to each, taking what he thought was good and rejecting what he disliked in their teachings. In religion, as in so many things, the Pacific Rim nations have always been willing to adopt new ideas—but they used them in their own ways.

More About Buddhism

At 4 A.M. the muffled tones of the morning gong awaken 16-year old Thong from his dreams. By sunrise he completes the formal ceremonies of candle lighting and recitation and is on his way to town. Dressed in the white robes of the *novice*, or beginner, monk and carrying a plain wooden bowl, he takes his place along a main street and waits for the city to come to life. Soon people stop by with food and other offerings to support Thong's local Buddhist monastery.

Thong's way of life is not unusual for some teenagers in his part of the world. In Asian countries such as Thailand where Buddhism is the religion of choice, young men are encouraged to put aside the ordinary concerns of daily life, at least for a while, and pursue a monastic life. Their purpose is to learn how to achieve salvation.

To Buddhists, salvation is a state of pure happiness. According to Buddhism, all human suffering is caused by people wanting what they don't have. To achieve happiness, or salvation, Buddhists believe one must rise above these desires. Such feelings cannot be stopped. But people can overcome them through discipline—study, *meditation*, or pure concentration, or following formal rituals.

Buddhists believe that trying to satisfy these desires is fruitless. More desires will simply arise to take their place. According to Buddhism, not only desires, but everything that we can see, touch, smell, and hear are constantly changing—arising and disappearing again. Our attachment to these things is the reason for our unhappiness.

Buddhists believe that this constantly changing world is governed by the law of *Karma*, or the law of cause and effect. There is both good and bad karma, good and evil causes. People who are unable to achieve happiness continue living in the karma-governed world of birth and death. Doing good brings a person closer to perfect happiness and allows him or her to leave the cycle. Failing this, when a person dies, he or she is *reincarnated*, or returned to this life, as another human or as an animal. One's happiness in the next life depends on his or her karma in the previous one.

The San Agustin Church in the intramuros (within the walls) fortress area of Manila, Philippines. Roman Catholics make up about 85 percent of the Philippine population.

Thus, in countries like Japan, people have a different attitude toward religion than do most people in the United States, Canada, or Western Europe. Japanese will offer a Shinto (their form of animism) sacrifice to solve everyday problems. Most will be married in a Shinto ceremony. But for funerals, most Japanese choose Buddhist rites. And even though less than one percent of Japanese are Christian, many young people are being married in Christian ceremonies—because they like the celebrations. At the Christmas season, downtown Tokyo is decorated with wreathes and Christmas trees.

In multicultural nations, people observe the religious holidays of all their religions. In Singapore, both Muslim and Buddhist holidays are celebrated. In fact, businesses use calendars marked on one side with the Buddhist year and the other with the Muslim year.

Ancient and modern. The blend of old and new is most apparent in the cities of the Pacific Rim. Glass skyscrapers reach into the sky, and people eat lunch at McDonald's and Kentucky Fried Chicken franchises. (The number-one McDonald's in the world, in terms of sales, is in the capital of Taiwan.) But at home the people keep their own cultural heritage as well. A Japanese girl may listen to American rock tape while she practices the traditional Japanese art of flower arranging.

Ancient arts are preserved everywhere. In Indonesia the beautiful cloth called *batik* is much in demand. In Thailand, craftspeople use traditional designs to make textiles and jewelry for the world's markets. Throughout Southeast Asia, people choose between a night at the movies or going to the equally popular shadow-puppet theaters.

Though the Confucian ethic remains strong, many young people are following more independent lifestyles. In the cities of Hong Kong, Singapore, and Seoul, massive population growth has reduced the social pressures found in villages. Traditionally, when a couple married, they went to live in the home of the groom's parents. This is still the case for the first-born son and his wife. The young bride had to follow the orders of her mother-in-law. Many young couples live in their own apartment in the city but still obey the husband's parents.

The rapid rise in the standard of living. One of the most amazing things about the Asian Pacific Rim is the rapid rise in people's standard of living. A high standard of living may mean different things to different people, but it usually includes a number of things that can be measured.

A high standard of living often means high pay for work done. Countries sometimes measure this by figur-

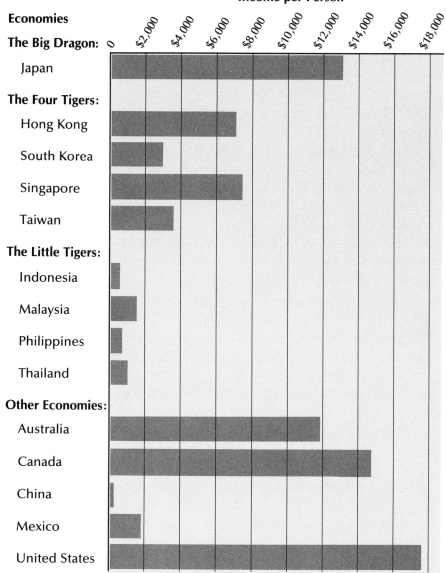

NOTE: Countries vary in how income per person is reported. The figure used is the Gross National Product (total value of goods and services produced) divided by the number of people as reported in the World Bank's *World Development Report 1988*.

ing out per capita income. That is the average income each person would get if all the income earned in a country is divided by the total number of people in a country. That figure is an average. In the past three decades per capita income in the United States has increased by about a third; in Singapore, it has more than tripled.

Another measure of standard of living is the quality of health care people get. One way to measure this is to find out how long the average person in a country will live. This is called life expectancy. The average person in Japan and Hong Kong, for example, now lives longer than does the average person in the United States. Another way experts measure health is to figure out the infant mortality rate. That is the number of babies who die for every 1,000 who live. Japan, Hong Kong, and Singapore now have a better infant mortality rate than the United States, since fewer babies die per 1,000 in these Pacific Rim economies.

Still another measure of a high standard of living is what percentage of adults can read and write—the literacy rate. Over the past 30 years, Pacific Rim countries have made rapid improvements in education. For example, in South Korea in 1960, about three in ten adults could not read and write. Today, fewer than one in ten cannot.

Progress, in the Pacific Rim as elsewhere in the world, is a two-edged sword. The price of land in Tokyo is higher than in any city in the United States, and the view of Mount Fuji is now often obscured by air pollution. A young Japanese politician warns, "If we get too rich, prosperous, and materialistic, all we do is create more unhappiness. We must therefore return to the true Japanese moral standards, including respect for our ancestors, compassion, gratitude, courage, cooperation, and obligation—all virtues neglected since the war."

The "war" he refers to is World War II, which created an upheaval in Asian society. Many of the Pacific Rim countries are politically less than 50 years independent, though their culture rests on centuries of tradition. In the past half century, they have gained wealth and power. They are still searching for ways to adjust to their new position in the world.

Improved health care means a happier life for Pacific Rim peoples, including these teenagers in the Philippines.

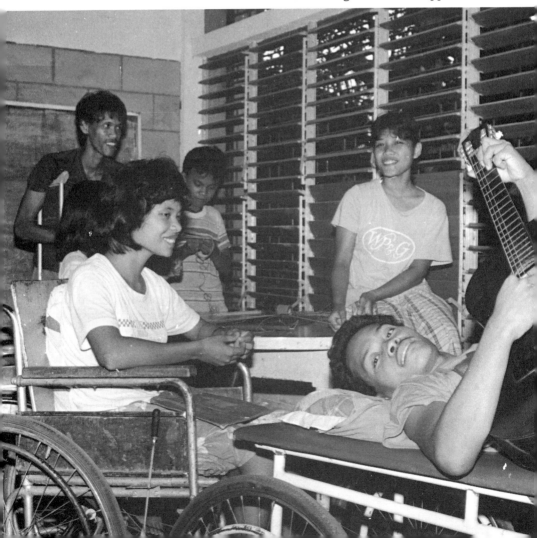

Double-check

Review

1. What is the meaning of the term *standard of living*?

2. What are four of the major languages spoken in the Asian Pacific Rim region?

3. Name four of the religions practiced in the Pacific Rim countries.

4. In what country did Buddhism begin?

5. How long ago did Confucius live?

Discussion

1. This chapter points out several Confucian values that have helped Asians in the Pacific Rim countries succeed. List these values. In what ways do you think these are similar to or different from the values you find in your community? Support your opinions.

2. Confucius's main concern was good government. He believed the most important job of government was to inspire people. Do you agree or disagree? Why? What do you think the most important goal of government should be?

3. A recent survey of how students performed in school in Asian countries showed that how well students did in school had little to do with how wealthy their families were. What have you learned in this chapter that might explain this?

Activities

1. A committee of students might be formed to prepare a large wall chart of the Pacific Rim countries. Students should use almanacs and other reference books to provide the following information for each country: major religions, main languages, per capita income, life expectancy, infant mortality rate, and literacy rate. After the chart is developed, students should compare and contrast the information for each country and reach several conclusions about life in each country.

2. Students could conduct a survey to find out attitudes toward schoolwork. They could ask their parents this question: How well would your child have to score on a test if the average score for the class was 70? Each student should write the answer on a sheet of paper without identifying who he or she is. The folded sheets of paper should be placed in a box. Tabulate the scores and find out what the parents considered a satisfactory grade. After the average is announced to the class, compare the findings with those of the survey mentioned on pages 39–40.

Skills

WHAT KIND OF WORK DO PEOPLE DO?

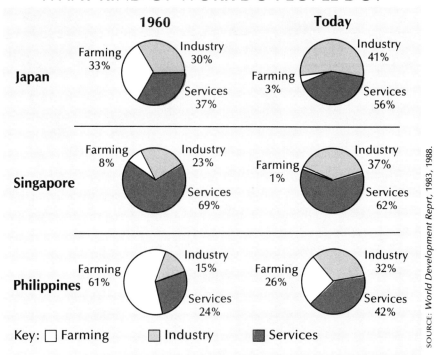

Key: ☐ Farming   ▥ Industry   ▨ Services

SOURCE: *World Development Reprt*, 1983, 1988.

Study the graphs. Then answer the following questions.

1. In 1960, which economy had the least part of its population working in agriculture?
 (a) Japan (b) Philippines (c) Singapore

2. In 1960, what was the percentage of people working in industry in Japan?
 (a) 9% (b) 30% (c) 49%

3. Which economy has the largest percentage of its workers working in the service industry today?
 (a) Philippines (b) Singapore (c) Japan

4. Which economy has the largest percentage of its workers having jobs in farming today?
 (a) Japan (b) Philippines (c) Singapore

5. Which economy's workforce most closely matches that of Japan in the percentage of people working in the service industry today?
 (a) Japan (b) Philippines (c) Singapore

Chapter 3

History

FOUR CENTURIES AGO, a single ship traveled across the Pacific each year. Only one ship—it was a Spanish galleon whose captain relied on the wind to fill its sails and a few crude instruments to find his way across 6,000 miles of empty ocean. Yet that ship carried the richest cargo of its time. Pirates lay in wait for it, and there were some years when they captured the Spanish treasure ship.

On the westward voyage, from Spanish colonies in Mexico, the ship's hold was filled with gold and silver. When it returned, it carried spices and silks that would bring a king's ransom in the markets of Europe. Spices such as cloves and pepper were grown only in the Orient, and they were truly worth their weight in gold.

To gain control of the spice trade, Columbus had crossed the Atlantic. When it became clear that America was not Asia, Ferdinand Magellan continued to try to

The bronze Daibatsu (Great Buddha) of Kamakura is one of Japan's great historical treasures.

reach the East by sailing west. In 1519, he circled the far southern tip of South America and headed across the Pacific.

Magellan was killed in the Philippines, but some of his men completed what was the first trip around the world. By the mid-1500s, Spain's *conquistadores** (conquerors) controlled the Philippines (named after King Philip II of Spain). From there, the Spanish carried on trade with the rest of the area we know as the Pacific Rim.

The westward trip to Asia was shortened drastically in 1914, when the United States completed the building of a canal across Panama in Central America. Ships from the eastern coast of the United States could now cross the Pacific in half the time. They no longer had to sail all around South America to get to Asia. President Theodore Roosevelt said, "The Pacific era, destined to be the greatest of all, is just at its dawn."

Few guessed how soon the Pacific era would reach high noon. Even in the early part of this century, the long trip across the ocean limited trade. Only in our time would cargo jets and worldwide telecommunications shrink space and time to make the nations of the Pacific Rim the United States' most important trading partners.

Colonialism. Spain was not the only European country to seek trade with Asia. Lured by the riches of the East, the ships of Portugal came first—followed by those of Spain, Holland, England, and France. By the end of the 19th century they had established colonies in all the Pacific Rim countries except Thailand and Japan.

Japan, which had long closed its ports to all but a few foreigners, saw four American steamships sail into Tokyo Bay in 1853. The American commander, Matthew Perry, gave a convincing demonstration of the power of his ships' cannons. The following year, the

A Japanese print showing American naval officer Matthew C. Perry visiting the city of Yokohama.

Japanese signed a trade pact with the United States. A new era began.

Later in the century, the United States became a colonial power like the nations of Europe. After a war with Spain in 1898, the United States took over control of the Philippines and other Pacific islands.

The colonial powers took much from their Asian colonies. In many places, they stripped the land of its natural resources. They sold their own products to the Asians and often taxed them for the privilege of being governed by Europeans.

But Europeans brought two important changes to the region. First, they introduced new crops that changed the economies of many Pacific Rim nations. The rubber tree was originally a South American plant. Britain and the Netherlands brought rubber trees to Malaysia and Indonesia and created huge plantations that are still the world's largest source of latex. The tea plant, originally found in China, was transplanted throughout parts of Asia where it had never been grown.

Europe also changed the populations of some of the Pacific Rim countries. Britain ruled the vast subconti-

COLONIAL POWERS IN THE PACIFIC RIM, 1914

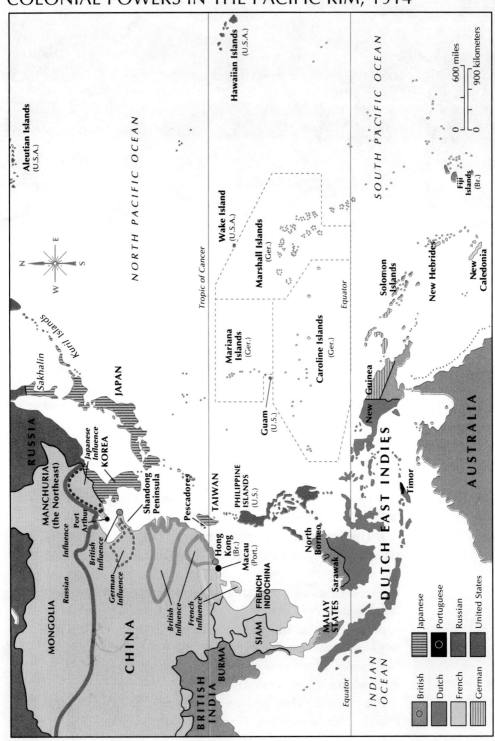

nent of India, and many Indians moved into British colonies in East Asia. They brought with them their culture and business skills. Today, Indian shopkeepers are found throughout Southeast Asian nations, such as Singapore.

World War II. While most of East Asia was being colonized, the Japanese studied Western ways and began to modernize. In particular, the Japanese found much of interest in Western guns and seapower. They strengthened their military forces and sought to carve out their own colonies. Short of natural resources and already feeling the pressure of a large and growing population, the Japanese turned on China. After a short war in 1895, they took the offshore island of Taiwan.

In 1905, the Japanese showed the world how much they had learned when they decisively beat the Russians in the Russo-Japanese War. The Japanese victory electrified Asia. It was the first time an Asian nation had ever defeated a European nation. It was a herald of things to come. In 1910, Japan added to its territories by annexing Korea.

Japanese troops moved into China's Northeast (Manchuria) in 1931. They invaded other provinces of China in 1937. These moves brought Japan into conflict with the United States. When diplomatic negotiations produced no agreement, the Japanese secretly decided to drive the American fleet out of the Pacific.

On December 7, 1941, Japanese planes bombed American ships anchored at Pearl Harbor in the mid-Pacific island of Hawaii. A day later, they invaded the Philippines. Rapidly, they took over Indonesia, Malaysia, and Burma. The British were confident that Singapore could never be taken by sea. But the Japanese thrust through the jungles to the north to take it by land.

Japanese troops invaded China's Northeast (Manchuria) in 1931.

By 1942, Japan was supreme in the western Pacific region. For the first time in history, these areas were brought under one rule. The Japanese called their new empire "The Greater East Asia Co-Prosperity Sphere." All the countries of today's Pacific Rim were under Japanese control except Australia and New Zealand.

The other Asians in the Co-Prosperity Sphere were treated cruelly by the Japanese. Japan exploited these areas as the Europeans had earlier. The memory of the Japanese occupation, and the hatred it caused, have endured among Asians to this day.

Post—World War II. After the defeat of Japan, the European powers tried to return to their old role in the colonies. But Japan's victories had destroyed the myth of European dominance. Throughout Asia, nationalism became a powerful force after World War II. By 1965, most of the Pacific Rim nations had won independence, either through armed revolt or by peaceful agreements.

However, struggle for control of the region was not over. After World War II, the United States and the Soviet Union competed for influence and control all over the world. This competition is known as the "cold war." It was a struggle between two economic ideals: capitalism and communism.

In Asia, the cold war flared into active fighting. The long Chinese civil war ended in a Communist victory in 1949. The remnants of the anti-Communist forces fled from the Chinese mainland to the offshore island of

National Day being celebrated in present-day Singapore. Singapore became a separate nation in 1965.

Taiwan. They set up a separate government of China there. Thus, mainland China and Taiwan were ruled by different governments.

Korea had been divided after World War II. Soviet troops in the north had set up a Communist government, while in the south, Syngman Rhee was elected president in an UN-supervised election in 1948. In 1950, North Korea invaded the south, and American troops, under the flag of the United Nations, went to the rescue. Three years of bloody fighting produced a stalemate that left the nation divided into North Korea and South Korea.

An American officer helps train South Korean troops to fight the invading North Koreans.

The United States moved to strengthen the Pacific Rim nations whose governments were sympathetic to capitalism. General Douglas MacArthur, leader of the American occupation forces in Japan, wrote a new democratic constitution for the country. The United States gave money and technology to help the Japanese and other friendly nations to build new industry.

At this time, none of the Pacific Rim countries were wealthy. Some ranked among the poorest nations of the world. Their main priority was to feed their people. The United States wanted to support these nations so that they could become bulwarks against communism, particularly against Communist China.

After the Japanese were defeated in World War II, the French returned to Southeast Asia, hoping to rule their colonies as they had before the war. The French were defeated by the Vietnamese in 1954. By the early 1960s, the United States began supporting the government of South Vietnam with economic and military help against Communist Vietnamese forces. Soon Communist North Vietnam and American-supported South Vietnam were involved in full-scale war.

By 1968 over 500,000 American troops were helping the South Vietnamese carry on the war. Despite massive bombings of North Vietnam by the Americans, the South Vietnamese were losing the war. In 1973, the United States, South Vietnam, Communist North Vietnam, and Communist forces in South Vietnam agreed to a cease-fire. The United States withdrew its troops.

Soon fighting broke out again between South Vietnam and the Communists. By 1975 Communist North Vietnam defeated the government of South Vietnam and united both parts of that country. Communists in neighboring Kampuchea (Cambodia) fought among themselves and with invading Communist Vietnamese. The

Saigon, now called Ho Chi Minh City. In 1975 Pro-American South Vietnamese attempt to flee from the victorious Communist Vietnamese by boarding an American helicopter.

result was millions of people killed by murder and starvation. Communist governments in Vietnam, Kampuchea, and Laos were unable to provide prosperity for their people.

Despite the American defeat in Vietnam, the American effort worked elsewhere, and the non-Communist Pacific Rim nations began to thrive. In 1967, a group of non-Communist nations in Southeast Asia formed an organization known as ASEAN (Association of South East Asian Nations). The members were Indonesia, Malaysia, the Philippines, Singapore, and Thailand. In the 1980s, the new nation of Brunei joined. ASEAN's purpose has been to encourage cooperation among members to stimulate economic growth.

The individual success stories of the Pacific Rim countries will be told in later chapters. But their success was in great part due to the help they received from the United States after World War II. In one respect, the United States "won" the cold war in Asia—most of the countries it backed stayed free from communism. Capitalism triumphed.

But it was a bittersweet victory. The Pacific Rim countries learned the lessons of economic development so well (they were, as Confucius wanted them to be, good students) that they now compete with their former patron and teacher. The fledgling nations that the United States fed with economic aid and technology now are our competitors in producing goods that the world wants to buy.

Double-check

Review

1. What European countries established colonies in the Pacific Rim countries?

2. What were some of the reasons why colonies were set up in Pacific Rim countries?

3. What Asian countries did Japan conquer from 1895 until the end of World War II?

4. When did the United States withdraw its troops from fighting in the Vietnam War?

5. What countries belong to the ASEAN organization? What is the purpose of that organization?

Discussion

1. List three wars that involved the United States in the Asian part of the Pacific Rim. Why did the United States fight in these wars? What was the outcome of each war for the Asian countries involved in them?

2. What changes did European powers bring to the Asian Pacific Rim counties? Were these changes beneficial to the people of the Asian Pacific Rim? Why, or why not?

3. What did President Theodore Roosevelt mean when he said, "The Pacific era, destined to be the greatest of all, is just at its dawn"?

Activities

1. Some students might do library research by reading news articles about Taiwan or Japan from forty years ago and from this year. They could prepare oral or written reports on the articles, emphasizing the changes that have occurred over the past four decades.

2. Some students might research and report to the rest of the class on the reaction of the Japanese to the Americans that arrived in Japan in 1853 and 1854.

3. Someone who fought in the United States armed forces during World War II, the Korean War, or the Vietnam War might be invited to speak to the class about his or her feelings about the peoples of the Asian Pacific Rim then and now.

Skills

PACIFIC RIM TIME LINE

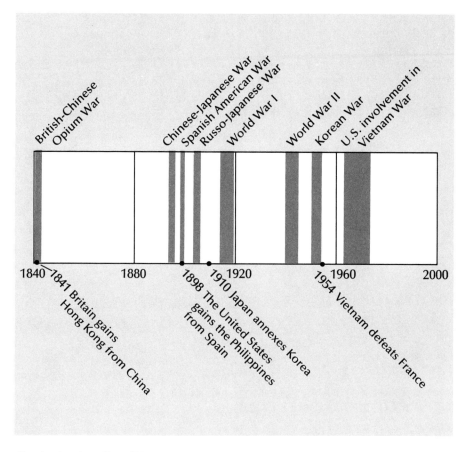

Study the time line. Then answer the following questions.

1. Which occurred first, World War II, the Russo-Japanese War, or the Vietnam War?

2. In what year did the United States take control of the Philippines from Spain?

3. When did Japan add Korea to its territories: before, during, or after World War II?

4. What country did the Vietnamese defeat in 1954?

5. About how many years ago did Britain make Hong Kong one of its colonies?

2
THE BIG DRAGON

Chapter 4

Japan: The Making of an Economic Miracle

According to an ancient Japanese legend, the sun goddess Amaterasu* created the Japanese islands from drops of water that fell from her spear. The emperors of Japan were said to be her descendants. The Japanese call their nation Nippon, which means "the source of the sun." The people who lived there shared a sense that they were special, different from other people.

Throughout their history, the Japanese have made the most of nature's gifts. The four major Japanese islands—Honshu,* or "mainland," Hokkaido,* Kyushu,* and Shikoku*—are together smaller than California, but are home to 123 million people. Less than

The old and the new in Tokyo. A Shinto temple near the entrance to the subway. Throughout Japan there are many historic sites that link modern Japanese to the past.

Winter in Sapporo on Japan's northernmost large island of Hokkaido. This city of 1.5 million people has snow for several months of the year.

one-fifth of that small amount of land is suitable for farming. Yet the Japanese have carefully used the land to make it produce as much food as possible.

Japan's islands are abundant in natural beauty. Mountains dominate the landscape. Japan contains more than one-tenth of all the active volcanoes in the world. Small earthquakes are frequent. About every ten years or so, there is one that causes much property damage and loss of life. In order to survive in this environment, the Japanese have learned to adapt to sudden disaster and change.

A tradition of borrowing. As an island people separated by 110 miles of water from mainland Asia, the Japanese remained independent from their neighbors. The Japanese were in a fortunate position—they could pick and choose what they wanted from China and from the other cultures of Asia.

In the sixth century, Buddhism was introduced into Japan. In order to find out more about the religion, Japanese missions went to China. They brought back the Chinese characters for writing. The Japanese found much to admire in the teachings of Confucius as well. They valued his ideas of family respect and of the duties people owed to superiors. The Japanese even copied the capital city of China. They built a duplicate of it for their emperor—today's city of Kyoto.*

A Japanese painting of a mountain house. Much of Japanese culture shows Chinese influence.

Although the Japanese borrowed much from Chinese culture, they also adapted it to their own ways of living. They practiced Buddhism alongside the Japanese form of animism known as Shinto.* The Japanese believed that spirits called *kami* lived all around them—in rocks, waterfalls, and the bodies of animals.

The emperor reigned in splendor at his court, but in the countryside, he had little power. Strong feudal lords divided the country among themselves, keeping order with their own bands of warriors, called *samurai*.* Samurai were fanatically loyal to their lords, living by a code of conduct known as *bushido*.* Bushido demanded a simple and strict way of life, skill at swordsmanship and archery, and a complete indifference to death. A samurai who violated the code knew, without being commanded, that he must kill himself.

In 1543, Portuguese ships arrived in Japan to open trade, and soon Roman Catholic missionaries from Spain and Portugal followed. Many Japanese converted to the new religion of Christianity. Even among those who did not, some wore rosaries and decorated their homes with Christian symbols.

A closed country. Among the things that the Portuguese brought to Japan were firearms. These new weapons appeared at a crucial moment in Japan's history. The most powerful of the feudal lords had claimed the title of *shogun*,* or "commander of the army." At this time, Japan was torn by a long civil war among those who wanted the title of shogun. The forces of Lord Tokugawa Ieyasu* won the war with the help of firearms. In 1603, he took the title shogun and ruled the country from his castle at Edo.* The Tokugawa family ruled Japan for 250 years.

The Tokugawas distrusted European influence on their country. They did not want to fall under the power of a European country the way the Philippines had been taken over by Spain. The Tokugawas outlawed Christianity and closed Japan to foreigners, except for one small port on the island of Deshima in Nagasaki* Bay. There, one Dutch ship a year was allowed to bring goods from Europe. Japanese were forbidden to leave the country.

The Tokugawas ruled with an an iron fist. Society was frozen into samurai, farmer, artisan, and merchant classes. At the top of society were the samurai. Only samurai could carry swords, and they could cut down without punishment people of the lower classes. The shoguns' government regulated such things as the sizes of houses members of each class could own and the clothing they could wear. The government even decided what kinds of toys the children of each class could play with.

Japan prospered under the Tokugawas. By the 1700s, Edo had grown to a city of over one million people. It became the largest in the world. The street in Edo where the country's mint, or *gin*,* stood, became known as the Ginza.* (Real estate there is today the most expensive in the world.) Roads were built throughout the country, and the town of Osaka* grew into a great commercial center. The city of Kyoto remained the home of the emperor. Though he carried out only ceremonial duties, Japanese gave him the respect owed to a descendant of the gods.

For 250 years, Japan remained closed to the rest of the world. Its isolation ended in 1853, when U.S. warships commanded by Commodore Matthew C. Perry steamed into Edo harbor. The United States wanted Japan to promise better treatment of any American sailors who might be shipwrecked on Japanese shores. The United States also hoped that Japan would provide harbors where American ships could stop for food and water. American steamships also needed a place to take on coal on their routes between California and Shanghai. Commodore Perry asked the Japanese to open their ports to American ships. He said he would give them time to consider his requests and would return "with a much larger force."

The Japanese had never seen steamships like those Perry commanded—they called them "the black ships." When Perry and his fleet returned the next year, the Japanese agreed to open two of their ports to trade. Perry gave the Japanese presents, including a model train and steam engine. The Japanese examined these and realized with a jolt that they had something to learn from other countries. Within a few years, the United States and several European countries signed trade treaties with Japan.

The "black ships" had exposed the weakness of the Tokugawa shoguns, and in 1868, the shogunate was abolished. The emperor moved his court from the inland city of Kyoto to the city of Edo, which was on the Pacific coast. Edo was renamed Tokyo ("eastern capital"). Japan was ready to redirect its efforts and power toward the sea and the outside world. Because the emperor took the name Meiji* ("enlightened rule") for his reign, this event was called the Meiji Restoration.

Japan's first economic miracle. The year 1868 is a very important one for Japan. In that year the Meiji era (1868–1912) began. The country saw an explosive growth in its economy, as well as drastic changes in Japanese society. The emperor expressed his hope for the country in a poem.

> *May our country*
> *Taking what is good,*
> *And rejecting what is bad,*

*Be not inferior
To any other.*

Japan's Meiji leaders recognized that modernization of industry would succeed only with modernization of education and the political and legal system. Just as one thousand years earlier students had been sent to China, now they were sent to Europe. There they were supposed to learn whatever the Japanese leaders saw Europeans doing most successfully.

Japanese delegations visited Western nations and brought back many suggestions. The emperor's new officials, who were mostly samurai, wrote a constitution. It provided for a lawmaking body called the Diet, a new tax system, and a legal code. The Japanese organized a public education system and established Tokyo University to train the best of Japan's young people. In a break with the past, a modern army was created, based on conscription. This is a system in which people are drafted into military service. Conscription ended the samurai monopoly on bearing arms since all young men were expected to serve. The samurais' treasured right of carrying a sword and their traditional way of dress were outlawed.

The Japanese saw that to compete in the modern world, they had to build a modern economy. In the 1860s, most businesses were small. Manufacturers produced handicrafts such as products made of silk, wood, porcelain, and metal. These were made in small workshops. There were no private businesses in Japan rich enough to build the necessary railroads and factories, so the government financed them. Soon, however, it sold the fledgling industries at bargain prices to private interests. These sales formed the basis of vast industrial concerns known in Japan as *zaibatsu.** These huge companies combined a wide variety of

Busy Yokohama waterfront in 1870.

businesses—banking, shipping, trading, mining, manufacturing, and others. The two largest zaibatsu were Mitsui and Mitsubishi. They trained their employees in new technology and guaranteed them lifetime jobs.

Japanese representatives went abroad to buy industrial machinery. Soon, however, they copied the machines and began building their own. In addition, they asked a small number of foreign experts to come to Japan. Since Japan lacked raw materials, such as iron ore, needed for modern industry, they imported them. This meant that the Japanese had to make goods to sell abroad. Then they could use the money earned by selling these exports to buy the foreign resources they needed.

Japanese leaders resisted foreigners who wanted to invest money to build factories and businesses in Japan. Instead, the government taxed the farmers very heavily. The tax money was used to build new industries.

One of the first new industries was textiles. For hundreds of years the Japanese had made fine silk garments. But now they used foreign machinery to weave cotton. The Japanese applied their high standards of beauty and quality to the products they made. Japanese laborers worked for wages far lower than those in Western countries. By the end of the 19th century, Japan was exporting its low-priced cotton goods to other Asian nations. And in the West, beautiful Japanese silk was much in demand.

Besides some coal in northern Kyushu, Japan had almost none of the natural resources needed to develop modern industry. It soon began to import the needed resources, especially iron to make steel. People moved from the villages and farmlands to the coasts. The Japanese cities along the coasts grew rapidly. Japan looked more and more seaward and outward to the rest of the world—a pattern that became even stronger in the 20th century.

Japan saw the economic benefits the West had gained from colonies. Many of Japan's leaders felt that it too needed an empire. Colonies would provide raw materials for Japanese industries and new markets for Japanese products. They believed that colonies would provide new land for Japan's large population. Just as European nations were doing, the Japanese began a drive to build their own colonial empire in East Asia. Their new army, equipped with modern weapons, took the island of Taiwan from China in a war fought from 1894 to 1895.

Ten years later, Japan defeated Russia in another short war. This was the first time an Asian nation had ever defeated a European one. Japan gained control of some of Russia's economic influence in part of China's Northeast (which was called Manchuria). Besides having an important railroad, the Northeast was and is rich in coal, iron, and other metals. It also produces abundant wheat, soybeans, barley, and corn. Japan had also taken from Russia control of the important harbor of Port Arthur (Lüshun) in China's Northeast.

With both China and Russia out of the picture, Japan took over Korea in 1910. It now could control Korea's resources, businesses, and land.

During World War I (1914–18), Japan sided with Britain, France, and the United States against Germany. When Germany lost the war, Japan gained control of Germany's island empire in the Pacific—the Marshall and Caroline Islands, and all the Marianas except Guam, which was a U.S. possession. During the wars it fought between 1894 and 1918, Japan diversified its industries. The result was that Japan could soon make a wider variety of products—mainly to help it win wars.

In 1922, Japan signed treaties agreeing to limit its battleships and aircraft carriers in a ratio of three to every five built by Britain and every five built by the United States. In effect, Japan promised to keep its navy smaller than Britain's or the United States' navies. However, Japan needed its ships only in the Pacific. Britain had to spread its slightly larger navy around the world to defend its empire. The United States needed ships in both the Atlantic and the Pacific. That meant that just 69 years after Commodore Perry's ships steamed into Edo harbor Japan had become the strongest naval power in the western Pacific. By the time Hirohito became emperor in 1926, Japan was recognized as one of the world's great powers.

Japanese naval power was put on display in 1926 when Hirohito became the new emperor.

More About Shinto

Shinto priests of the Toshogu Shrine walk through the rain typical of Nikko, a city north of Tokyo.

Imagine that you are enrolled in a Japanese high school. Along with students throughout Japan, you are about to take a test that will determine your chances for getting into a top university. Your parents are understandably anxious for you to do well. Like many other parents at this time of year, they visit a Shinto shrine in Tokyo that is dedicated to the god of learning. There, they write a prayer requesting the god to help you succeed. They place their prayer on an *ema,** or wooden tablet, which they hang on a tree at the shrine. As they leave, they feel reassured that they have placed you under the guardianship of the spirit of learning.

Such scenes are common in Japan. At many important turning points in their lives, the Japanese turn to an ancient native religion known as Shinto and make requests of its deities. At such times, they renew their sense of belonging to a unique cultural heritage, their spiritual roots.

What does it mean to practice Shinto? The answer is more complex than you might think. In contrast to Christianity, which can be understood as a set of beliefs and moral rules, Shinto has no sacred text or guide to moral behavior. Instead, Shinto in it purest sense is a way of life. This way celebrates nature and humans' harmony with the natural world. Every tree, mountain, bird—indeed, every natural object—has its own *kami,** or spirit. The Japanese have built shrines throughout their islands to pay homage to thousands of spirits, or gods. These shrines vary greatly in design and setting. Some are great temples in the midst of parks. Others are tiny altars set up in out of the way streets in Tokyo. At these shrines the Japanese believe that they can feel the presence of spirits of nature, of their own personal household gods, and of their ancestors and the spirits of the departed. The Japanese believe that their entire culture should be lived in harmony with the kami in nature and in the invisible world. The name Shinto comes from two Chinese words meaning "the way of the gods."

Shinto's origins are ancient. For centuries before the arrival of Buddhism and Confucianism from China into Japan in the sixth century, the Japanese had been following "the way of the gods." Shinto was so deeply embedded in

Japanese culture that it was not even thought of as a religion. In fact the word *Shinto* did not exist until the Japanese needed to distinguish their beliefs from those of the Chinese. "The way of the gods" was as much a part of Japanese life as breathing the air and eating rice.

If they converted to Buddhism or followed the teaching of Confucius, the Japanese did not give up their old way of life. They did not stop going to Shinto temples to pay homage to their ancestors and the kami they say are in nature. Instead, they practiced Shintoism, Buddhism, and Confucianism side by side. The Japanese saw no conflict in belonging to several religions or traditions at once. Some Japanese scholars believe that the Japanese took aspects of Confucianism for their law, Buddhism for their relationship to the world at large and the eternal, and Shinto to aid them in their daily life. Over time, all these religions and traditional ways adopted aspects of the others. Shinto joined together traditions taken from Chinese and Korean, as well as Japanese, sources.

By the 19th century, Shinto had adopted Confucianism's emphasis on loyalty to the emperor, to the family, and to the nation. Indeed, the emperor himself was regarded as a visible Shinto god and chief priest. In 1868, the Meiji government decided to use Shinto to enforce obedience to the emperor and to his policies. The government declared that Shinto was a state religion. It decreed that Shinto be taught in schools, and that all Japanese belong to a Shinto shrine.

Until the end of World War II, Shinto was used to justify Japan's aggressive, militaristic policy. If the emperor was a god, it was argued, surely anything he wanted to do was right. Opposition to his plans was sacrilegious. By the end of World War II, the emperor's position as a god had come to symbolize Japan's militaristic past. Therefore, the occupying American army made Emperor Hirohito renounce his deity and declare publicly to the Japanese nation that he was a human being without divine powers.

Today, Shinto is a private religious organization. In the absence of state sponsorship, it has returned to its roots.

Many Japanese bring rice and flowers to Shinto shrines and write their prayers on emi.

Shinto is still an important force in Japanese politics and culture. When Emperor Hirohito died in 1989, there was an intense international debate. Should his son, Akihito, take part in the Shinto *dai josai*,* the rice ceremony? During this ceremony, the son of the dead emperor becomes a god as well as an emperor. Some Western governments thought that to follow the ceremony might revive memories of Shinto's association with militarism. However, most observers believe that Akihito is thoroughly Westernized and has no desire to return to the old imperialist Japan.

A debate that may have more far-reaching effects centers around Japan's tradition of growing rice. In 1988, the United States asked Japan to end its ban against imported rice. This policy has protected small rice farms that have existed in Japan for centuries. The United States wants Japan to buy more American rice. Unexpectedly, the Japanese government reacted angrily that rice growing "is closely related to the Japanese culture."

The government insists that rice growers are living in communion with nature because rice is the Shinto "staff of life," much as bread is to Christians. "Rice is our Christmas tree," said Hideaki Kase, an expert on Shinto ritual. *Manichi,* a national daily newspaper, added: "Americans think that if they shove, Japan will give in, but now they've stepped on sacred ground."

The trade debate demonstrates that even in the 20th century, the Japanese feel strong ties to Shinto. As they did over a millennium ago, they seek to live in harmony with the natural world and invisible spirits.

Double-check

Review

   1. What do the Japanese call their nation?

   2. Who were the samurai?

   3. List four changes brought about during the Meiji era.

   4. Why did large cities develop along Japan's coasts instead of in the center of each large island?

   5. What role did the zaibatsu play in the Japanese economy?

Discussion

   1. During the late 19th century the United States allowed wealthy people in Britain and other countries to invest in American businesses. Why did Japan resist such foreign investment when it, too, was developing its industries in the late 19th century? Which path toward development do you think is better? Why?

   2. How did Japan hope to benefit from owning colonies? Do you think it is possible to get such benefits for a country by means other than building an empire of colonies? Support your opinion.

   3. What three changes in Japan do you think were most important in making Japan a world power in just a few decades? Do you think these kinds of changes would make a developing country a world power if tried today? Why?

Activities

   1. Some students might draw political cartoons or posters depicting Commodore Perry's arrival in Edo harbor.

   2. Some students might research and report on the popular religious practices and celebrations in Japan today. They should be sure to include a presentation of the role of Shinto in modern Japanese life.

   3. Students should form two committees to do research on several of the zaibatsu. One committee should make a presentation to the class pointing out the benefits of such an economic organization. The other committee should present the disadvantages. The committees should then elect a chairperson to lead a classroom discussion and then to conduct a vote favoring or opposing zaibatsu.

Skills

TIME LINE: JAPAN BECOMES A WORLD POWER

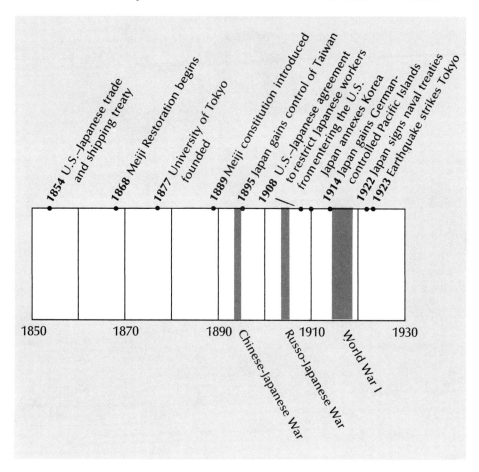

Use the time line to answer the following questions.

1. After which war did Japan gain control of Taiwan?

2. In what year was Tokyo University founded?

3. When did the Meiji leaders give the country a constitution?

4. How many years passed between the Japanese-American agreement allowing ships into Japanese ports and the agreement forbidding Japanese to move to the United States?

5. In what year did a violent earthquake strike the Tokyo area?

Chapter 5

From Military Disaster to Economic Triumph

IN THE 60 YEARS FROM THE 1920s to the 1980s Japan underwent a series of rapid changes militarily, politically, and economically. In the 1920s, young Japanese lived in a Japan very different from that of their grandparents.

Rapid change had unsettling effects on Japanese society in the 1920s and 1930s. Many people were distressed by the clash between Japanese traditions and the modern values associated with the West.

Change and strain. Let us see how modernization produced strains and conflicts in Japanese society.

1. The growth of big business. The huge business companies known as zaibatsu continued to grow. Many

*Cars are readied for export
from the port city of Nagoya.*

Japanese resented the concentration of great wealth in the hands of a small group of zaibatsu families. They felt that these business magnates had too much influence, power, and prestige.

2. Population growth. After 1868 there had been a dramatic increase in population, largely because of modern medicine. By 1925, the population had doubled from 30 million to 60 million. It was increasing at the rate of one million a year.

Despite Japan's industrial growth, the number of jobs could not keep pace with the number of people. In rural areas, where half the people lived, "surplus" young men migrated to the cities in search of work. Often they found that there was none, or that wages were very low. They became a great reservoir of "cheap labor."

Many people who remained on the farms were no better off. By 1920 almost half of Japan's farmers were reduced by poverty to the status of tenants. They worked on land owned by others and paid the rent with a share of the food they produced.

3. The growth of cities and mass culture. Japan's cities began to look more and more like those of Western Europe and the United States. This was especially true after 1923, when an earthquake and fire destroyed half of Tokyo. To prevent another such disaster, the city was rebuilt with many steel and concrete structures. Wide avenues replaced once-narrow streets in the downtown area. Soon other cities began to emulate Tokyo with modern office buildings, schools, movie houses, stadiums, and railroad stations. By this time Tokyo had more than two million people, and Osaka had well over one million.

As Western architecture made its appearance in Japanese cities, so did the mass culture of the West. During the 1920s many young girls in the United States began to wear short dresses and skirts, and also cut their

86

In 1923 an earthquake destroyed much of Tokyo. One of the buildings that remained standing was the Imperial Hotel (in the upper righthand corner of this photo) designed by American architect Frank Lloyd Wright.

hair short. They were called "flappers." Many city girls in Japan copied these styles. They were called *moga*,* short for the English words "modern girl." Their boyfriends were known as *mobo*,* short for "modern boy." American movies, jazz, and styles of dancing became quite popular, especially among the young. There were other signs of American mass culture—musical shows with chorus girls, dance halls, and bars. America's national pastime, baseball, became Japan's national sport, too. Other Western sports, like tennis, track and field, and swimming, became almost as popular as baseball.

Western social customs also made some headway among young people in the cities. Traditionally, Japanese parents acted as matchmakers for their children and arranged marriages for them. Parents believed that young people were too impractical to choose suitable marriage partners themselves. Now, however, some young people insisted on making their own choices,

87

guided solely by love. Traditionally, too, Japanese girls were taught to become obedient, hardworking housewives. That was a duty they were supposed to owe their husbands. But now many women worked in offices, and some were beginning to question whether women should be little more than household servants.

The Western life-style that was developing in the cities bewildered and offended many Japanese. Older people were shocked by the behavior of the "modern" girls and boys. In the rural areas, which were hardly touched by Western influences, most people still believed in the old values—family solidarity, obedience to authority, duties, and obligations. Distrust and disapproval of city ways were common. This attitude was shared by most of Japan's army and navy officers. They believed that the cities were corrupting the young and making them "soft." The cities were a threat to Japan's warrior traditions of toughness, self-discipline, and loyalty to superiors. Among these conservative groups and others there was a longing for the past, for a return to a simpler, more harmonious, and more authoritarian Japan.

4. The growth of democracy. By the 1920s Japan developed a democratic system of government that outwardly was similar to that of Great Britain. Democracy in Japan seemed to be secure, but in reality it was threatened by a number of weaknesses. Under the Meiji constitution, powerful officials, including the leaders of the armed forces, could claim that they acted for the emperor and were responsible only to him. They were above control by elected party politicians. Beyond this, most Japanese had no deep commitment to democracy. Many of them, in fact, distrusted it. There was a widespread belief that Japan's major political parties were corrupt and were controlled by the zaibatsu business combines.

In the 1930s, students took part in military drilling. Why did militarism grow in Japan in the 1920s and 1930s?

Militarism takes over. The worldwide depression of the 1930s crippled Japan's economy. Its exports fell because other nations had imposed high tariffs, taxes on goods brought into their countries. Many factory workers lost their jobs. Hard times set in. Extremist nationalist groups believed that the solution to the problem was to expand Japan's empire. At first both the government and the zaibatsu were opposed to military conquest. They believed that Japan's economic security and well-being depended on peaceful trade and international cooperation.

Military leaders and Japanese patriotic societies continued to urge the government to obtain more land through military conquest. The military had many supporters among the farmers, who made up the bulk of army and navy conscripts. For most farmers, service in the armed forced was an exciting alternative to the drudgery of working in the rice fields. In school they had been taught reverence for the emperor and the nation, and the "glory" of dying for them in combat. Many of these conscripts were thrilled to follow in the footsteps of the samurai nobility of old.

In 1931, the Japanese army took control of China's Northeast (Manchuria). Japanese settlers went there to exploit its rich natural resources. The easy conquest of such a huge area, much larger than Japan, made more and more Japanese very nationalistic.

Japanese expansion in north China led to a full-scale war with the Chinese in 1937. Japan's armies won victory after victory, but the Chinese forces retreated inland and continued their resistance. The war in China dragged on, taking a heavy toll on Japan's resources.

At home, Japan's military leaders tightened their control over the government and repressed political opposition. Hundreds of left-wing politicians, labor leaders, and university students were thrown into prison and tortured until they renounced their views. In 1940 the government forced all political parties to dissolve.

The outbreak of World War II in Europe in 1939 created new opportunities for Japan's military leaders. They proclaimed their intention of creating a vast empire in Asia. They called their grand design the Greater East Asia Co-Prosperity Sphere.

After German armies crushed France in 1940, Japan's leaders signed a treaty of alliance with Europe's Fascist partners, Germany and Italy. Japanese troops also began moving into Indochina, which was then a French colony.

The United States had done little more than make futile diplomatic protests against Japanese aggression. It was selling to Japan large amounts of scrap iron, oil, and other raw materials, all of which were essential to the Japanese war machine. Finally, in 1941, after Japan took over Indochina completely, President Roosevelt shut off all U.S. shipments of raw materials to Japan, including oil.

Without U.S. oil, Japan's industry and armed forces would grind to a halt within a year. The leaders of Japan were now faced with a critical dilemma. In order to buy American oil, they were being asked to give up their conquests in China and Indochina. This they were unwilling to do. The other alternative was to launch a war to seize the Dutch East Indies (Indonesia) where oil was plentiful.

Japanese military disaster. The Japanese gambled that they could beat the United States in a short war. On December 7, 1941, planes took off from Japanese aircraft carriers to bomb the American fleet at Pearl Harbor in Hawaii. The attack destroyed 90 percent of all U.S. naval and air forces in the Pacific. Now

Emperor Hirohito on his favorite white horse.

virtually unopposed, the Japanese attacked and overran much of East and Southeast Asia in the next five months. They told the conquered countries that they were being freed from European colonial rule. Their Greater East Asia Co-Prosperity Sphere would result in an "Asia for the Asians." But the Japanese treated both war prisoners and the civilian populations of other Asian countries with great cruelty.

The gamble failed. The attack drew the United States into World War II. American industry turned its full power to the goal of winning the war. Factories stopped making cars and refrigerators and began to make ships, planes, and tanks. Soon, new American ships were in the Pacific, and a U.S. fleet defeated the Japanese at the Battle of Midway in 1942. U.S. Marines took Japanese-held islands one by one.

As American forces moved closer to Japan, the Japanese prepared for an all-out defense of their home islands, which no invader had ever conquered. American planes bombed Tokyo many times, creating fires that destroyed much of the city. On August 6, 1945, the United States dropped an atomic bomb on the Japanese city of Hiroshima. Three days later Nagasaki was also hit by an atomic bomb.

Some still did not want to surrender, but Emperor Hirohito overruled his advisers. On August 15, 1945, he spoke to his people on the radio. It was the first time they had ever heard his voice, and many knelt before their radios. He called on them to "endure the unendurable." On September 2, the formal surrender was signed on the battleship *Missouri* in Tokyo Bay. Japan was stripped of all its conquests.

An American shogun. In September 1945, one could walk for miles in Japan and not see a complete building. Two million Japanese had been killed. Most of the survivors were homeless and hungry. They waited with apprehension for the Americans to occupy their country. Having no experience of defeat, they felt spiritually desolated. The people who had been the favorites of the gods now awaited a conqueror.

The man in charge of the occupation was General Douglas MacArthur. He ruled postwar Japan like an American shogun. MacArthur had grown up in the Philippines, where his father was an American military governor. He knew Asia better than most westerners. He first had to decide what to do about the emperor. Many Americans wanted him tried as a war criminal. When Hirohito made his first formal visit to MacArthur, he took responsibility for the war: "It was done in my name, and I am responsible." But MacArthur decided that the emperor could keep his throne provided that he renounced his divine status.

MacArthur had a new constitution written for Japan. It renounced war forever and abolished the military forces. The emperor was described as the "symbol of the state and of the unity of the people." He would have no more power than the British monarch. Japan's people would elect a parliament, and a prime minister would head the government. For the first time, women were given the right to vote.

Under MacArthur's direction, Japan's society changed. Many of the zaibatsu were broken up. In the countryside, an ambitious land reform program distributed land to the farmers who worked it.

American aid programs sent food and other supplies, and gradually conditions improved. Much of Japan's industry had been destroyed by the wartime bombing, but its people again began to make export goods.

A big boost to Japan's economy came during the Korean War (1950–53). In that war, U.S. troops supported South Korean forces fighting against the invading

North Koreans who, in turn, were supported by Communist China. Japan's harbors were used as bases for American ships. The United States began to look on the democratic nation of Japan as an Asian bulwark against Communist aggression. It supplied Japan with technology, aid, and advice, for it wanted Japan to be a strong ally.

The second economic miracle. Through the sacrifice and hard work of its people, Japan started to move ahead. The slogan for development was, "heavy, thick, long, big." Heavy industry, such as steel and shipbuilding, was stressed. The destruction of its factories during World War II turned into an advantage. In rebuilding almost everything, Japan could create an entirely modern industrial base. By 1957, Japan had the most efficient steelworks in the world. These advances in basic industries laid the groundwork for Japan's export industry.

The pace of economic rebirth picked up throughout the 1960s. The first public notice of Japan's development came at the 1964 Olympics, held in Tokyo. Television carried pictures of a bright, shining city, with new skyscrapers and modern hotels. Visitors were treated to a ride on the new "bullet train" from Osaka to Tokyo—the fastest and most modern in the world.

Japan began to export automobiles, TV sets, and cameras. Each year in the 1960s, the Japanese economy grew by more than 10 percent—a phenomenal rate. Although workers' pay did not keep pace, it too inched upward. Japanese were able to buy home appliances and small luxuries for themselves.

In the 1970s the growth rate slowed. Because Japan was totally dependent on foreign oil, the quadrupling of oil prices drastically increased their manufacturing costs. But they cut their consumption of oil, and their fuel-

Why did the Japanese concentrate on making transistorized radios, TVs, and computer parts in the 1970s?

efficient cars became more popular abroad. As more Americans bought Japanese cars, Japan started to have a trade surplus with the United States. For the first time, it was selling more to the United States than it bought from the United States. The other side of the coin was that the United States now had a trade deficit with Japan. At first, few Americans seemed to think this was a problem. But by the 1980s, the imbalance of trade was so great that it caused serious tensions.

The oil crisis of the 1970s caused the Japanese to change their industrial priorities. The new slogan was "light, thin, short, small." Electronic products required less energy to use and to make. Thus, Japan concentrated on making transistorized radios, televisions, and computer parts. New jobs opened up in electronics

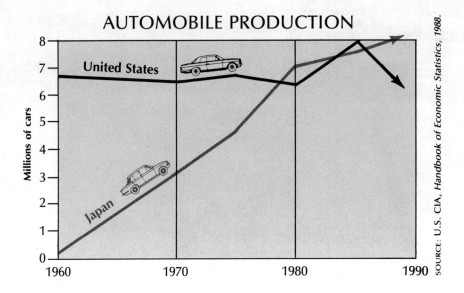

SOURCE: U.S. CIA, *Handbook of Economic Statistics*, 1988.

plants, which had better working conditions than heavy industry did.

The Japanese skill in miniaturization helped them make smaller, lighter electronics products. Within a few years, Japan gained an enormous share of the consumer electronics industry. The United States virtually abandoned manufacturing its own TV sets. The production of computer chips became another Japanese success story.

By the mid-1980s, Japan was moving into its next stage of development—called "beauty, feeling, play, creativity." Tokyo, like Paris and Rome, is now a center of high fashion. Each spring, buyers from the world's stores come to see the latest Tokyo lines.

Leisure industries were developed, such as computer games. In computers, the emphasis shifted from making hardware to developing software programs and researching high-tech developments such as new kinds of computer chips. The United States had long dominated the field of high technology, but now its lead was threatened.

By the 1980s, the economic slogan was "beauty, feeling, play, creativity." Japan concentrated on high-fashion, computer games, high-tech software, and new kinds of computer chips.

Economic triumph. Forty years after its defeat, Japan had the most dynamic economy in the world. It had succeeded by peaceful means in gaining the prosperity it had failed to achieve by war. It was pumping out 10 percent of the world's output of goods, and its economy was the third largest in size. Only the United States and the Soviet Union had larger economies. Japan's products were regarded worldwide as high in quality. Their wealth from trade surpluses was overflowing. The ten largest banks in the world were Japanese. The value of the stocks on Tokyo's Stock Exchange was higher than those on Wall Street. Tokyo now was the control center of Japan's fabulous wealth, which was being invested throughout the world. Japan was seen as Number One and its economy was being copied by others.

Double-check

Review

1. What were three of the ways rapid change affected the Japanese in the 1920s?

2. Define the terms *moga* and *mobo,* and explain how they represented change.

3. What was the Greater East Asia Co-Prosperity Sphere?

4. What does the Japanese slogan "light, thin, short, small" mean?

5. List three industries in which Japan has surpassed the United States in the past thirty years.

Discussion

1. Because of its economic miracle, Japan now sells more to the United States than the United States sells to Japan. Do you think this is a problem for the United States? for Japan? Why? Should either government try to do anything about the trade imbalance? Support your opinion.

2. What are some of the lessons of history that other countries might learn from the Japanese experience during and after World War II? Explain.

3. Japanese students learn very little about Japan's militarism of the 1930s or about Japan's role in World War II. A Japanese government office makes sure that Japanese textbooks do not discuss these issues in depth. Do you think this lack of knowledge about their nation's past is harmful or beneficial to today's Japanese students?

Activities

1. Japanese banks and other businesses have branches in the United States. A Japanese businessperson might be invited to speak to a class about the Japanese business system and how his or her company has changed over the past twenty years.

2. Several students might collect newspaper and magazine articles about Japanese business expansion and create a bulletin-board display to provide current information for the class. The display might be based on several themes: Japanese Consumer Goods, Japanese High-Tech Industries, Japanese Banks, Japanese Investments in the United States.

3. Students might make lists of the Japanese-made products found in their homes. Then the students might compile a list of the three most popular Japanese products.

THE HEIGHT OF THE JAPANESE EXPANSION

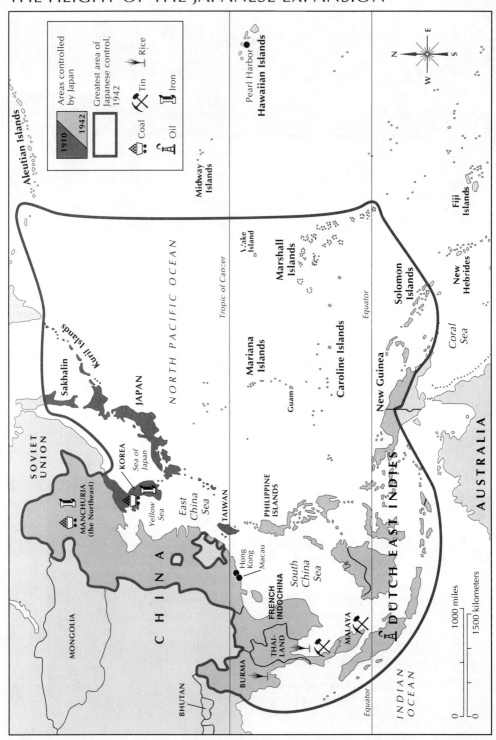

Skills

READING A MAP

Study the map on page 100. Then answer these questions.

1. From which countries in its Greater East Asia Co-Prosperity Sphere did Japan obtain tin?
    a. Korea and Taiwan   b. Mongolia and the Soviet Union   c. Thailand and Malaya

2. Where did Japan obtain the oil it needed for its industries, navy, and army?
    a. Dutch East Indies   b. Soviet Union   c. Marshall Islands

3. What important resources was Japan able to obtain from China's Northeast
    a. rice and wheat   b. coal and iron   c. oil and tin

4. What valuable item did Burma and Thailand provide for Japan's large population?
    a. coal   b. oil   c. rice

5. In what year was the Japanese colonial empire at it largest?
    (a) 1910   (b) 1941   (c) 1942

Chapter 6

Living in an Economic Miracle

"IN JAPAN," SAYS WRITER ICHIRO Kawasaki, "work is a ceremony." The workplace is the center of life for Japanese. The Japanese term *Sha-in,** means "company man." It stands for member. Each worker is encouraged to feel that he or she is a member of a work community, not just an employee. It adds to the sense that being an employee is like being a member of a family. Understanding the Japanese at work—in the office, in the factory, and in school—is crucial for understanding what it is like to live today in Japan's economic miracle.

A harmonious workplace. The attempt to foster a family feeling in the workplace goes on from morning to evening. Workers begin the day by singing the company song or taking part in an exercise routine. Throughout the day, *jishu kanri,** or "voluntary management

Measured by income, education, and health care, the Japanese now have one of the highest standards of living in the world.

Workers at a Nagasaki shipbuilding yard do their daily exercises.

groups," meet often with the workers to discuss problems and ask for suggestions on how to increase productivity. Workers' ideas are given respectful attention. Middle managers regularly come to the factory floor to talk with workers. The boss in Japan is a friend and father-figure and often gives his workers advice on their personal problems at home.

After hours, workers and managers often go out together for drinks and dinner. This socializing, called *nemawashi** or "rootbinding," strengthens the bond between workers and bosses. It gives them a chance to discuss ideas in an informal setting without the fear of "losing face" by publicly making an unpopular or mistaken suggestion. Many companies regard *nemawashi* as a business expense, with managers picking up the bill. This business socializing means male workers usually do not have dinner with their families. They may get home from work at 10 P.M or midnight on weekdays. For many people, Saturdays are also workdays.

The emphasis on group decision making fits traditional Japanese values. An old proverb says, "If a nail sticks up, you hammer it down." Since most unions are organized by company rather than industry, the union member's loyalty is also to the company. The larger companies in return have offered lifetime employment and often offer low-cost homes, free medical care, and lessons in such subjects as English and flower arranging.

Men dominate the workplace. Women were expected to be full-time housewives and mothers even if they worked before marriage. In recent years, however, women have been coming into the work force in great numbers. Today, around 40 percent of the work force

Electronics workers. Do Japanese women have opportunities equal to those of men in the work force?

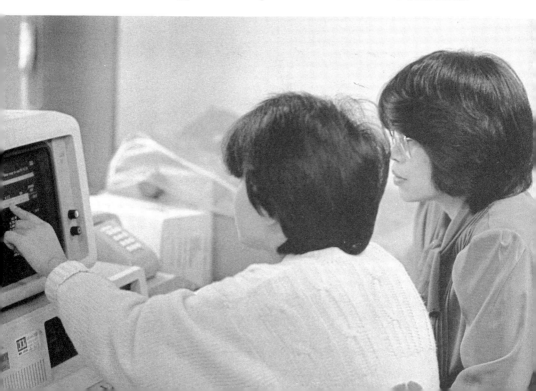

are women, but they make little more than half the pay of men in comparable jobs. And they have little chance for advancement to the managerial level.

Business executives are the modern-day samurai with their devoted loyalty to the company and a firm sense of responsibility to the workers. Prospective executives usually are put through a rigorous training program. Top executives have a great deal of hands-on experience. Konosuke Matsushita, head of a huge electronics firm, claimed, "I have worked at every job in the shop: on the assembly lines, in the accounting offices—even emptying trash cans. I can believe in the value of practical

An executive in a high-tech business. Many Japanese work six days a week. However, starting in 1989, banks and some other companies were closed on Saturdays, giving many Japanese workers two-day weekends for the first time.

experience. You can analyze chemically the taste of sugar—even write a book about it—but nothing beats taking a bite."

When something goes wrong in Japanese corporations, the top executives take full responsibility, often by making public apologies. Even when the problems are not really their fault, executives will give themselves big pay cuts to show their shame. The standard apology includes stating their sorrow and taking "responsibility for disturbing society." For example, after a plane crash, the president of the Japanese airlines personally apologized to the victims' families and then resigned from his post. When one of Japan's largest companies was involved in a scandal, its president bowed his head in apology on a nationwide TV broadcast to employees. He also cut his own pay by 20 percent although he wasn't suspected of involvement in the scandal. "Japanese executives bear much broader responsibilities than Americans," says economist Keitaro Hasegawa. "Japanese society isn't satisfied unless top executives accept responsibility when something happens. It's a time-honored Japanese practice."

Six keys to Japan's economic miracle. What accounts for the great success of Japan? There are six main reasons.

- *American aid.* Teams of experts came to help rebuild Japan's shattered economy after World War II. The United States also helped Japan to win favorable trade agreements with other countries in the early postwar years.

- *Government-business partnership.* Japan followed a policy of protectionism to shield its industries from foreign competition. It set import taxes, or tariffs, which raised the prices of foreign goods. Sometimes,

THE ECONOMY OF JAPAN

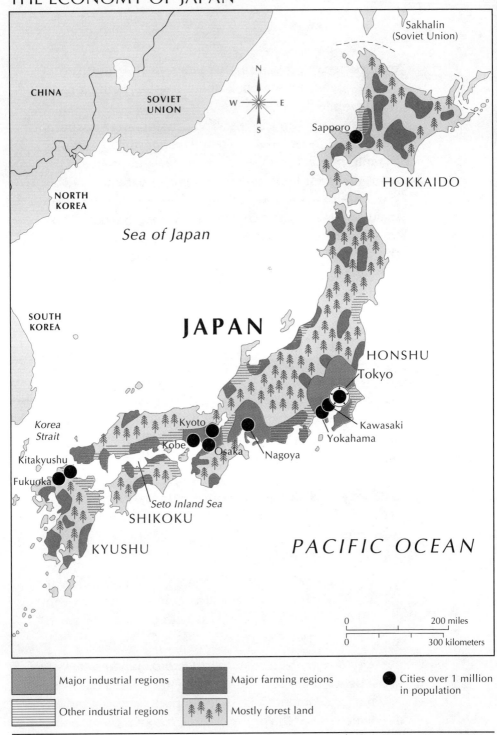

108

quotas limited the number of imported goods. The U.S. occupation established the first trade barriers to help Japan compete. But as Japan's wealth grew, protectionism became a sore point between Japan and its trading partners.

The government-business partnership helped businesses buy technology rights for which they paid relatively little. The government built railroads and ports, rather than creating parks or helping with consumer economy. This created a strong infrastructure— the foundation or base— on which the Japanese economy could develop.

- *Low defense budget.* Even after they were allowed by the Unted States a self-defense force, the Japanese did not spend more than one percent of their national budget on the military. (Before World War II, Japan had spent 17 percent on defense.) This freed money for productive economic development. Japan still depends on the United States for its defense. In addition, Japan could use its best scientific minds to develop consumer goods.

- *Low cost of Japanese goods.* At the beginning of its post–World War II economic growth, Japan's workers earned much less than their Western counterparts did. Prices were also kept down by the low value of the Japanese yen. The United States had set the Japanese currency at a very low value, which made Japanese products cheap on the world market—often even cheaper than they were in Japan. Other savings came from efficient equipment in new factories. Japanese businessmen were willing to accept a low profit per item in order to gain a greater share of the market.

- *High rate of personal savings.* In 1985, the average

Japanese household had over the years saved about $28,000—more than their annual earnings. In most years Japanese workers save one-sixth to one-fifth of the money they've earned, after paying taxes. (In contrast Americans on average, save about one twentieth, and in recent years even less.) Since Japan did not have easy credit plans to encourage consumers to borrow, they had to save to buy such things as refrigerators or houses. These savings generated pools of cash that banks used to finance continued business expansion.

- *High quality of workers.* Japan has the most highly educated work force in the world, and the workers are willing to work long hours.

A strategy for success. The strategic planners of Japan's dynamic economy work within a special department of the government. The Ministry of International Trade and Industry (MITI) is in charge of coordinating the country's industrial policy. Working closely with business, MITI gives guidance on how Japan's resources should be allocated, that is, divided and used. It helps companies get loans from banks and suggests which infant industries should be protected by discouraging imports. Businesses can then "put to sleep" those industries deemed unlikely to be competitive on the world market and convert to new, more productive, industries.

There is no counterpart to MITI in the United States. Perhaps because of the rich natural resources in the United States, national economic planning has not been an important government role. But in Japan, conditioned by its scarce resources, planning is regarded as essential. For example, Japan now leads the world in exporting high-tech machine tools and factory robots because it planned ahead to surpass the United States and West Germany in this field. Members of MITI study the world to spot Japanese trade opportunities and early trouble signs for Japanese business. The country's sense of vulnerability and the feeling that mistakes could be catastrophic still survive. MITI members are as important in Japan as the high-ranking members of the National Security Council are in the United States.

MITI's services give Japan a national economic strategy that the United States lacks. It is felt that if large businesses made their decisions independently, then precious resources might be wasted and the national good might not be served. Capitalism in Japan is different from the American brand. American capitalism offers opportunities for riches for the individual that will in turn enrich the country as a whole. In Japan the good of the country comes first and many believe that the best arbiter is a government servant. The high prestige of serving MITI is shown by the term *amakudari**—"descended from heaven." That is what former MITI members say they have done when they leave government service to enter the business world.

Quality, quality, quality. The products of many Japanese corporations are respected the world over. For example, Akio Morita, the chairman and cofounder of Sony, was asked why his electronic products were so popular with the American consumer. "Many reasons," he replied, "but I would say the first three are quality, quality, quality. Many Americans think that is something new to Japanese industry, but quality is an age-old tradition among us."

For centuries, Japanese crafts workers and artisans have had high standards. They took pride in their products, whether they were made for the imperial court or for country markets. This tradition of high quality and

The tradition in Japan is to save a large part of one's earnings and to pay for goods in cash. Only in the past several years has borrowing money by the use of credit cards begun to spread, especially among younger shoppers.

attention to detail is one of the great strengths of modern Japan.

The service in Japan reflects this as well. Shoppers entering department stores are greeted by young women whose job is to bow and bid them welcome. Restaurant windows display plastic replicas of the items on their menus that are virtually works of art. A car pulling into a gas station is immediately surrounded by teams of workers who check over the engine, clean the windows, and fill the gas tank while the driver rests in a comfortable waiting room inside.

Some doubted that the high standards of skilled artisans could be applied to mass production. But the Japanese were greatly influenced by the ideas of an American, Dr. W. Edwards Deming, who believed that by following correct procedures, quality could be built into the product. Deming's system of quality control consisted of five steps:

- Add up defects.
- Define them.
- Trace them to their sources.
- Make corrections.
- Record the results.

Deming had offered his ideas to American businesses but was rejected. He was invited to Japan to address managers on quality control. Deming promised that if they followed his rules they would become competitive with the West in five years. A listener recalls, "Here was this tall, strange American telling us that we would be an important force in five years [1955] if we did what he said. We really didn't believe him, but in order not to lose face, we did what we were told and it worked."

In gratitude to Deming, Emperor Hirohito awarded him a medal. Each year the Union of Japanese Scientists and Engineers recognizes the firm that most advances quality control. This award is called the Deming Prize.

The key that winds the watch. "The cherry blossoms are falling" is a message that brings heartbreak to many Japanese students and their families. These words mean that the student has failed the entrance examination for Tokyo University. Those who take the test put in hundreds of hours of study and give up sports and other amusements to cram day and night for a year. Many people feel the effort is worth it, for acceptance at Tokyo University is considered almost a guarantee of lifelong success.

Japan's passion for education is one of the most important reasons for the country's success today. Fully 99 percent of the population is literate. The historian Frank Gibney calls Japan's educational system "the key that winds the watch."

Japan's schools make heavy demands on their students. In order to be literate, a student must memorize and learn to write 1,850 characters needed to read a newspaper. Through the years, additional characters must be mastered for more serious reading.

Because classes are large, students are taught to pay close attention to their teacher. In the younger grades, the children cooperate to keep the school clean—there are no janitors. Lunch is served in the classroom by the students. No one is left back, and the handicapped are taught with the rest of the students.

From preschool on, the school year is two months longer than in the United States. School is also held a half day on Saturdays. The curriculum is the same throughout the country, and there is a heavy emphasis on math and science. Education is compulsory through junior high. To get into senior high, the student must pass an entrance examination.

Japanese parents want their children to get into the best schools. Students at these schools have a built-in advantage when taking the tests for the next level. They are called "escalator schools" because they put their students on the road to success. For the very best schools, there are entrance exams even at the kindergarten level.

Students start studying for senior high school exams over a year in advance. Tutors may be called to the home or parents take the job of overseeing their children's studying. Special cram schools prepare students. A typical student will spend additional evening hours several days a week at the *juku*,* or cram school. A favorite saying is, "Six hours sleep fail, four hours sleep pass." The process, repeated for high school finals, has been called Japan's examination hell because it puts enormous pressure on the students.

The pressure to avoid the disgrace of failure is great.

Teenagers studying in their room. Why do the Japanese consider education so important?

Another pressure for teenagers is bullying at school. Bullying, or *ijime,** is carried out by the "in groups" against students who they feel do not measure up. Junko Katohda is a high school senior who lives in Mikita City, a suburb of Tokyo. She describes how difficult it can be if a student is seen as different. "The Japanese have a strong sense of group," she says. "I have my own individuality and creativity, but I refrain from displaying it. If you deviate from the group, bad things could happen."

Junko's home is typical of many Japanese homes today. It is a mixture of traditional Japanese and modern. The family eat fried chicken, but, in traditional Japanese fashion, they sit on the floor while eating. Her grandmother might perform the traditional tea ceremony before dinner in a special room reserved for the ancient ritual. Junko is trying to learn the complex steps of the ceremony. But like other Japanese teenagers,

Junko likes many American things—particularly American ice cream. She wants to mix Japanese traditions with modern culture.

Junko wants to go into the business world and become a secretary. She is studying to get into college. In contrast to many of her contemporaries, she tries to study at her own pace. "You hear a lot about how much pressure there is on students," Junko says. "But that's only if you want to go to a 'super A-class' school."

Looking to the future. The death of Hirohito in 1989 marked the end of a tumultuous era in Japanese history. Outside Japan it revived memories of Japanese brutality in World War II. Within Japan it provoked discussion about the incredible changes the country has seen. And it raised the question of what role Japan should play in the world—how it should use its new wealth. Japan's success has come so rapidly that the Japanese themselves have not adjusted to it. This was a time to evaluate what they have become.

The Japanese realize that their wealth has made other countries envy, as well as admire, them. Japan's position of economic leadership has not been matched in other fields. Ironically, the feeling of Japanese specialness that has brought them success in business has hampered them in international dealing. They have been accused of not shouldering the responsibilities of wealth. Japan has tried to improve its image by increasing its aid to poorer countries. In 1989, it passed the United States as the largest donor of foreign aid.

The relationship with the United States is Japan's most important one. The United States has been Japan's conqueror, big brother, protector, and finally its best customer. Japan's huge trade surplus has put strains on the relationship, as has its reluctance to open its markets to foreign products. Japan has used its wealth to make large investments in the United States. Japanese bond-buyers have helped finance part of the enormous American budget deficit. The two countries are closely bound together in a relationship that many in both countries have criticized as unequal.

Japan is being pressed by the Four Tigers—South Korea, Taiwan, Singapore, and Hong Kong—whom some see as "hungrier" than Japan. Already, the Tigers are competing with Japan in steel, shipbuilding, computers, and cars. They have the advantage of lower labor costs. In fact, Japanese firms have moved some of their factories to these countries, causing fears among workers at home. To remain on top, Japan must keep its lead in high technology.

This new Asian competition comes at a time when Japan is facing internal changes. Almost 60 percent of its economy is in the service sector—one in which workers provide services rather than making goods. In addition, with the longest life expectancy in the world, the Japanese face the problems of an aging population. Traditionally, the elderly are cared for in the home. But medical care and other services for them may become a drag on the economy of the future.

One problem that cannot be solved is lack of space. An average Tokyo family of four lives in what would be a one-bedroom apartment in the United States. As Tokyo has grown, land has become more expensive. In the late 1980s, property on the Ginza in downtown Tokyo cost $28,000 per square foot, the most expensive in the world! Skyscrapers reach ever higher. There is less parkland than in most cities. But there is no end in sight as more people and businesses move in all the time.

Since World War II, Japan has developed a society in which there has been a remarkable degree of equality of income. Almost everybody belonged to the large middle class. In the past ten years, however, a major difference has grown between those who own homes and those who don't. Land prices have skyrocketed. Those who already own homes are considered well off. Those who now wish to buy a house must go heavily into debt and have little money left for consumer goods and education costs.

Besides very high prices for housing, the Japanese also pay high prices for everyday consumer goods. Because of the way the tax system is set up, Japanese goods are often cheaper abroad than they are within Japan. Even many well-to-do Japanese families therefore do without appliances such as clothes dryers. By the late 1980s, about half of the Japanese questioned in opinion polls expressed dissatisfaction with some aspects of Japanese life. These included high prices, long workdays, and long commuting hours (often two hours each way.) There were also complaints that the government has spent very little on building an adequate sewage system. Nor are there enough parks and other places for recreation.

At present, Japan seems to have an identity crisis. Many older Japanese fear that today's young people lack their work ethic. They look at teenagers in their blue jeans and T-shirts enjoying the latest fads, and wonder whether they will be willing to sacrifice the way their elders did. They worry about opinion polls that show more than half of the younger Japanese do not want to work in one company all their lives, and that they want to enjoy their leisure time.

Japanese in their 50s and 60s built an economic miracle in a generation, and they want the younger generation to build it further. These older Japanese may identify themselves with the Japanese spirit, or *Yamato damashii*.* They call today's youth the "bean sprout generation"—growing fast, but weak. Every Sunday, groups of young Japanese dress up like American rock-

Young people gather in Tokyo's Yoyogi Park to dance to rock-and-roll and wear 1950s-style clothes.

and-rollers of the 1950s and flock to Yoyogi Park in Tokyo to dance. Elders look on with disapproval. They fear that today's young people are dedicated to pleasure rather than duty.

Though young people agree that their spirit of sacrifice is not so strong, they argue that it is no longer needed. "Young people in Japan today have no desire to work as hard as our parents," says Junko Katohda. "Our parents are workaholics. We're willing to work, but we want entertainment, too. We want to go out, listen to music, and travel." These desires will influence Japan's future in the 21st century.

Double-check

Review

1. List three ways the Japanese attempt to create a harmonious workplace.

2. Why are business executives sometimes called the modern-day samurai?

3. List four of the keys to Japan's economic miracle.

4. Why do so many Japanese students attend *juku,* or cram school, in the evenings?

5. What are three important things that MITI, Japan's Ministry of International Trade and Industry, does to help Japan's economy grow?

Discussion

1. Japan has been very successful in taking scientific discoveries and turning them into successful and profitable products. One area where Japan is weak is in pure science—the kind that has no immediate use. Japan's regimented schools do not encourage questions from students. Japan's labs do not encourage risky research that is not aimed at an immediate goal. Japanese leaders are worried that Japan does not seem to be overtaking the United States in pure science and this failure may mean remaining second to the United States in high technology based on science. How might Japan benefit by setting up an American-style school system?

2. In what ways might your daily life be improved if Deming's system of quality control, used in Japan, were put into practice?

3. If other countries used Japan's methods of achieving economic change do you think they could be equally successful? Support your opinion.

Activities

1. A committee of students might research the Japanese school system by using library resources and recent magazine and newspaper articles. Half the committee members should present to the class the major strengths of the Japanese school system and half should present its weaknesses. A classroom discussion might follow the presentations.

2. Some students might draw cartoons illustrating the daily life of a worker or student in Japan.

3. Several students might role-play a meeting of MITI held to discuss and find solutions to one or more of the following problems facing Japan: long working hours, long commuting hours, high cost of consumer products and appliances, high cost of houses, lack of enough parks.

Skills

WHAT AVERAGE HOUSEHOLDS SPEND THEIR MONEY ON

Economy	Percentage Share of Total Household Consumption					
	Food	Clothing and Footware	Medical Care	Education	Cars	Other
Hong Kong	19%	9%	6%	5%	1%	60%
Japan	19%	6%	10%	7%	1%	57%
United States	13%	6%	14%	8%	5%	54%

NOTE: The amount spent is the percentage of household income after taxes are paid.
SOURCE: *World Development Report, 1988.*

Use the chart to answer the following questions.

1. In which country does the average household spend the largest percentage of its income on clothing and footwear?
    (a) Hong Kong (b) Japan (c) the United States

2. In which country does the average household spend the largest percentage of its income on medical care?
    (a) Hong Kong (b) Japan (c) the United States

3. The average household in Japan spends what percentage of its income on education?
    (a) 5% (b) 7% (c) 19%

4. If your family were an average household in Japan, it would spend about how many yen on food for every 100 yen earned, after paying taxes?
    (a) 10 out of every 100 yen (b) 13 out of every 100 yen (c) 19 out of every 100 yen

5. In which country does the average household spend the largest percentage of its income on autos?
    (a) Hong Kong (b) Japan (c) the United States

3
THE FOUR TIGERS

Chapter 7

South Korea: The Miracle on the Han

A CENTURY BEFORE Columbus reached America, a Korean general, Yi Song-gye,* overthrew the Korean king. Believing that the current capital was not grand enough, General Yi decided to build a new one. He sent advisers known as geomancers to find the most favorable location. After studying the energies of earth, water, and air, they chose a place in the Han River valley. Thousands of workers came to build palaces and temples. Called the Fortress on the Han, General Yi's city is today known as Seoul* (capital).

General Yi would be proud of his city. Home to almost one-fourth of South Korea's 45 million people, Seoul is the fifth largest city in the world. Skyscrapers, including Asia's tallest, stand where there were once thatched houses. It is the capital of a country with one of the fastest-growing economies in the world, the center of what is called "the miracle on the Han."

In early spring, South Koreans use Yoido Plaza, in the captial city of Seoul, to roller-skate and bike.

THE KOREAN PENINSULA

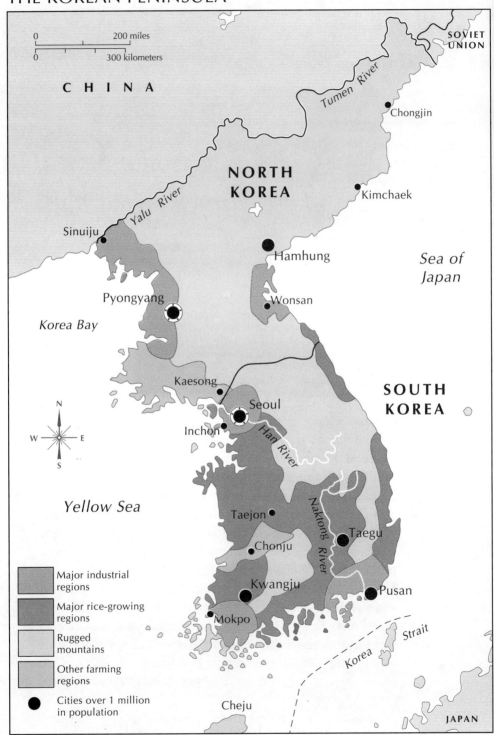

124

A shrimp between whales. Koreans say their land "is like a shrimp caught between two hungry whales." Throughout its history, Korea has been influenced by its two larger neighbors, China and Japan. From China, Koreans received Buddhism, Confucianism, and the Chinese written language. In peaceful times, Korea has acted as a cultural bridge between China and Japan. Buddhism, for example, spread first to Korea, and Koreans brought it to Japan. But in times of war, Korea has been a battleground in which its powerful neighbors fought for dominance.

Although Koreans learned much from Chinese civilization, they developed their own unique culture. Koreans are proud of the fact that they invented moveable type 200 years before Gutenberg did in the West. And, after using the Chinese script, the Koreans invented an alphabet, called Hangul,* with 11 vowels and 28 consonants. To ward off a Japanese invasion in the late 16th century, a Korean admiral invented the first ironclad ships—called "turtle ships."

After that invasion, the Koreans turned inward. Called the "hermit kingdom," Korea shut off contacts with all countries except China. But by the 19th century, the Hermit Kingdom was too weak to remain aloof from world affairs. Korea's strategic location on the mainland of Asia caused China, Russia, and Japan to compete for influence there. After Japan defeated the other two countries, it annexed Korea in 1910.

For the next 35 years, Japanese settlers and *zaibatsu* exploited Korea. The Japanese tried to wipe out Koreans' national identity. In the schools, all teaching was in Japanese. It became illegal to speak Korean even at home. Koreans had to take new, Japanese-style names. Japan destroyed many of Korea's cultural treasures.

The Korean War. The Japanese occupation of Korea ended with Japan's defeat in World War II. But Korea was divided at the 38th parallel. Soviet troops took the surrender of Japanese troops north of that line, and American forces moved into the south. The division was supposed to be temporary. But it hardened as hostilities increased between the United States and the Soviet Union. In the North, a Communist dictatorship under Kim Il Sung was established, and Syngman Rhee led the government of the South. In 1948, both superpowers left, and two hostile Korean governments faced each other across the 38th parallel.

On June 25, 1950, North Korea invaded South Korea in an attempt to reunite the country by force. The United States immediately sent troops to defend the South. Soon, the United Nations voted to sponsor a military force to halt North Korea's aggression. But the North's forces had moved swiftly, taking Seoul and pushing the South Koreans and their allies into the far southern tip of the Korean peninsula.

General MacArthur of the United States led the United Nations troops. He staged a brilliant amphibious landing behind North Korean lines at Inchon. The North Koreans were driven back across the 38th parallel. MacArthur pursued, pushing his troops north until the North Koreans were backed up against the Chinese border.

Then, the Communist government of China entered the war on the North Korean side. China sent a million-man army over the border and overwhelmed MacArthur's forces. Bloody fighting forced the United Nations troops into a long retreat back across the 38th parallel. Seoul once more fell. The fighting seesawed back and forth until a truce was signed in 1953. It set a border between North and South at roughly the same place it had been before. American troops remained in South Korea to guarantee its safety.

President Syngman Rhee faced an enormous task of

Rice harvesters. In 1965 over half of all workers were farmers. Today, fewer than one out of four workers has a job in agriculture.

reconstruction. Seoul had changed hands four times during the fighting and lay in ruins. Refugees wandered the country, their homes destroyed by the war. In the early postwar years, South Korea survived on foreign aid from the United Nations and the United States.

Joint ventures and *jaebul*. President Rhee was forced to resign because of widespread protests against his strong rule. In 1961, an army officer, Park Chung Hee, seized power. Park was a military dictator, but he set in motion an incredible economic recovery.

Although South Korea had few natural resources,

Park believed that an export economy was the road to success. He gambled that Korea could attract foreign investment money. He could offer investors Korean workers, who were diligent, worked long hours for low wages, and were forbidden to strike. These conditions brought investment from both American and Japanese businesses. They built factories in Korea as joint ventures with the Korean government. South Korea benefited by gaining modern technology.

In the first stage of industrial development, Korea made textiles and simple electronic items like calculators. By exporting these products, Korea earned money to help lessen its debts. Results started to show by the mid-1960s. In the 10 years from 1962 to 1971, exports increased by a phenomenal rate of 40 percent a year.

With the technological knowledge they gained from foreign-run industry, Koreans began to develop their own businesses. A major role was played by huge family-owned businesses called *jaebul*. Modeled after the Japanese *zaibatsu,* the *jaebul* operated in many fields of manufacturing and finance. Their size and government connections enabled them to dominate the Korean economy. These giant companies are able to invest billions of dollars in high-tech product development. The four largest are Hyundai, Daewo, Samsung, and Lucky Goldstar.

The *jaebul* sprang from nowhere to become worldwide conglomerates in less than 30 years. Daweoo began by trading in textiles in 1969. Today, Daewoo manufactures and sells computers, electronics products, and automobiles around the world. Chung Ju-Yung, today the head of Hyundai, was a poor farmboy who walked 300 miles to Seoul to seek his fortune.

As in Japan, the oil price rises of the 1970s hurt Korean economic development. But Korea pressed forward, moving into heavy industry—steel, shipbuilding, and automobiles. To modernize the nation, the government built roads, subways, and railroads. Even widespread political unrest and the assassination of President Park did not halt the economic surge.

   Koreans were tireless in building their nation's industry. They were inspired by the Japanese example, and they copied its governmental direction of industry. But they wanted to compete with, or even surpass, Japan. Rapidly, Korea moved from being an underdeveloped economy into a "NIE," or newly industrialized economy. South Korea is the largest in population of the Four Tigers. Although it also has a larger economy than do Taiwan, Hong Kong, and Singapore, its workers earn much less per person.

By the mid-1980s, new miles of subway had made Seoul's subway the seventh largest in the world.

Japanese shoppers in Tokyo buy South Korean VCRs, which cost them 20 percent less than Japanese ones. South Korean companies are eager to invest in new kinds of high-tech consumer electronics.

In the 1980s, Korea moved ahead thanks to "the three blessings"—cheap oil, low interest rates, and cheap money. The low value of South Korean currency made its products cheaper in foreign markets. Now South Korea competed directly with Japan and the United States in making more sophisticated electronics products. Korean VCRs captured one-third of the world market. South Korea is also fast becoming a major producer of color TVs and microwave ovens. In computers, Koreans built lower-priced "clones" that duplicated the machines of such giants as IBM. In 1987 none of the most advanced computer memory chips were produced by South Korea. Yet by the end of 1988, only Japan, the United States, and West Germany produced more of these. Plans are now under way to compete with Japan and the United States in the fields of aerospace, and telecommunications equipment and robots for use in factories.

Koreans even succeeded in the highly competitive international market for automobiles. Hyundai produced a car called the Excel, with Japanese quality at a much lower price. When it was introduced into the United States in 1986-87, it became the fastest-selling import in American history. But as South Korea's success produced a trade imbalance with the United States, it caused a strain in relations between the two countries. By the end of the 1980s, the United States had become the world's largest debtor nation. In contrast, South Korea is fast becoming a creditor nation, with more owed to it than it owed others.

Seoul is where the action is. The 1988 Olympic Games, held in Seoul, let Koreans show off their shining

Fiber optics in production. South Korea is advancing rapidly in developing its technology.

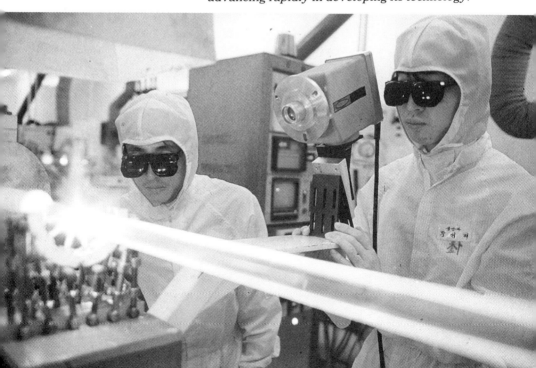

new capital and sports facilities. For a few weeks, the world's eyes shifted to Korea, and Koreans felt they were given the due their hard work had earned them.

South Korea's miracle economy is directed from Seoul, seat of the government and headquarters of most *jaebul*. The sound of jackhammers whirring is constant, as new buildings are going up. Seoul is a mixture of the very modern and the traditional. Broad new boulevards and narrow streets carry almost half of the cars in the country, making for big traffic jams.

Seoul has been a magnet for South Korea's population. People continue to come from the countryside, eager to find good jobs and share in Korea's new wealth. The city grows to accommodate them. South of the Han River, where rice paddies stood 20 years ago, there are now rows of high-rise apartment buildings. These are the homes or *apart*, for Korea's new middle-class.

North of the river is the older part of Seoul. Its more spacious homes are occupied by the very rich, but lower-class Korean factory workers live in crowded apartments on narrow streets. In its teeming blocks are also found small businesses where people labor long hours making clothing.

Seoul is famous for its markets. People browse among the open-air stalls, stopping to haggle over prices with the merchants. At Namdaemon Market, a steady stream of cars and buses moves around the Great Southern Gate of the original city. At the old east gate is Tongdaemon Market, the largest in Asia. Here, everything from the latest electronic gadgets to Korean folk art is sold. In addition, beneath some of the city streets there are modern shopping arcades.

Amid the skyscrapers, Seoul has many reminders of its past. The Korean government has restored General Yi's palace in the heart of the city. Residents can stroll through the Secret Garden, once a private park re-

A North Korean government official (left) visits the huge Lotte Department Store in downtown Seoul.

stricted to royalty. Here one can find a respite from the city's hubbub among the trees and shrubs, artificial ponds, and pavilions.

Even though Seoul pulses with its new economic strength, there are reminders that the Korean War has not officially ended. Once a month the city halts for an air raid drill. Seoul is less than 35 miles from the border with North Korea. Prosperity is a fragile thing in this proud Pacific Rim capital.

Work, school, and leisure. The Korean work ethic is second to none. The country has the longest work hours in the world—55 a week, on average. Sunday is the only full day off, although most businesses have a half-day on Saturday. Summer "vacation" is two days long. It is customary for employees to be at work when the boss arrives and not to leave until he does. In companies, including *jaebul*, junior executives salute their bosses by lowering their heads. Daewoo's leader, though now a very wealthy man, sets a stern example. "In a country with a $2,800 per capita GNP," he said, "we shouldn't play golf. Poor people can't afford that. I tell my executives, 'No Golf!'"

The pay of ordinary workers has not increased in proportion to the prosperity of the country. Moreover, in the rush to industrialize, safety standards have often been lax. These conditions have led to labor strife. For years unions were illegal—strikes brought violent fighting between workers and police. Even after the government legalized unions, company goon squads have kidnapped labor leaders and beaten them.

Education is a passion of the Koreans. Confucian respect for teachers and scholars is a part of their very long tradition. Even the head of Daewoo says, "Before I die, I want to see businessmen respected as much as professors."

In some ways, Korean students are more fortunate than the Japanese or Chinese, for the Hangul alphabet makes reading and writing easier to learn. On October 9, a national holiday celebrates its invention.

Just as in Japan, the right school is important for advancement later in life. To get into senior high schools and colleges, a student has to pass a rigorous examination. Science and math are strongly emphasized. Korean parents tell their children, *"Kongbu haera"*—Study! On their one day off, Sunday, students from 7th grade on will have an early breakfast and hit the books for 12 hours. The competition is made fiercer by the fact that each student can apply to only one school. Those who fail face a year of study before they can apply to another school.

The Korean economy suffers from a severe shortage of engineers for its high-tech industries. So students direct their studies to mathematics and science to prepare for such jobs.

Traditionally, Korean families decided whom their children would marry. Even today, a date between two young people is often arranged by a go-between. The English word *meeting* means a blind date to a Korean. The suffix *-ting* is added to other words that describe the activities of young romantics. *Go-ting* means "dancing," from disco's go-go dances, and a date outdoors such as picnicking is *ya-ting* after the Korean word *ya*, which means "outside."

Teenagers in Seoul relax on a Sunday afternoon. Most students spend their time studying on their one day a week off from school.

On Sunday, families often go on trips to the mountains outside the city. Koreans particularly enjoy group activities, such as hiking trips and neighborhood picnics. On Children's Day, a national holiday, all activities are free for children. Spectator sports such as baseball and soccer are widely followed. The streets are empty during major boxing bouts, as people watch on TV.

Korean soap operas, called "home dramas," are popular TV programs. Real families gather before their sets to watch the triumphs and tragedies of fictional ones. People follow the problems of TV families, so like their own—will the middle son get into college? What will grandma say about her granddaughter's new boyfriend?

Democracy and reunification. Korea's economic miracle took place under strong military rule, but as Koreans grew more prosperous, they wanted greater freedoms. Student-led protests against the government often resulted in all-out fighting between protestors and police.

In 1988, after a democratic election, Roh Tae Woo became president. President Roh gave workers the right to form unions and promised to allow peaceful dissent. The police and intelligence services, so repressive in earlier regimes, became less visible.

Reunification is the most important issue in South Korea today. North and South Koreans are ethnically one people. Many families were separated before and during the Korean War. The truce left about 10 million relatives stranded on either side of the 38th parallel.

President Roh has tried to smooth the way for what he called a "springtime for peace and reconciliation on the Korean peninsula." He has improved relations with the Soviet Union, the People's Republic of China, and Communist governments in Eastern Europe. South Korea has begun trade with these countries, both to find

Protestors demanding the arrest of former South Korean President Chun Doo Hwan, who ruled repressively. Mr. Chun had apologized for his harsh rule and went to live in a remote Buddhist monastery in the mountains in order to repent for his wrong-doing.

new markets and to ease tension with the Communist world of which North Korea is a part.

However, North Korea has carried out terrorist actions against the South. North Korea has also insisted that the 40,000 American troops leave South Korea. But in 1989 President George Bush spoke before the Korean parliament, and promised that the troops would remain.

In a break with the past, the head of Hyundai, Chung Ju-Yung, visited Pyongyang,* North Korea's capital. His talks with North Korean government officials resulted in an agreement for the joint development of a resort in the North. Direct negotiations between the two governments began in 1989. If South and North can someday re-unite, Korea will become an even more important force in the economy of the Pacific Rim.

More About South Korean Junior and Senior High School

These South Korean students are on a school tour to learn about painting and sculpture. This exhibit is in a subway station.

Imagine a schedule on which you study 12 hours a day, and during school vacations you go to school six days a week, sit in unheated classrooms in winter, and have no time to date.

That's what a secondary school is like in South Korea, says Sung Min Yoo, who ought to know. He graduated from high school there and is continuing his education in the United States.

"In high school it's very hard work... very stressful. We don't have much time to socialize. After school we usually stay at school under the supervision of a teacher and do some sort of study," he said.

Students in South Korea don't have to go to school under these conditions. They are required by law to attend only public elementary schools. The government cannot afford to pay for school beyond sixth grade. It can't even pay to heat some schools in the winter because the price of fuel is so high. But over 90 percent of South Korean teens attend junior and senior high school and pay tuition. Korean students are taught that education is the key to personal success and to prosperity for their country.

All ninth-grade students take exams for admission to high school. Depending upon their scores, students can enter an academic high school that will prepare them for college or a vocational high school. Vocational high schools train students for careers in agriculture, business, fishery, and marine industry, among other fields. Students whose scores are not high enough for admission to academic or vocational schools are assigned to general high schools in areas where they live.

Schoolwork in South Korean junior high and academic senior high schools is very difficult. Students get little or no individual help from teachers because classes contain as many as 60 students. They learn Korean and English languages, social studies, nature studies, math, chemistry, physics, calculus, scientific experimentation, art, and physical education.

The work is so demanding, South Korean students spend more time in school than American students do. The school day in South Korea begins at 8 A.M. and ends at 5 P.M. After school, students work independently or join study groups. Classes meet six days a week with early dismissal at 1 P.M. on Saturday. Nearly all students go to school during their summer and winter vacations for review.

South Korean students have practically no social life. Their studies take most of their time and energy. They are not encouraged to date until they go to college or graduate from vocational high school.

"Dating in high school is not illegal or forbidden but students are looked down on if they do. Less than 10 percent date," says Sung Min.

The school system in South Korea is even set up to discourage dating. Most boys and girls go to different schools. In the small but growing number of co-educational public and private schools, boys and girls attend different classes. The scholastic standards have their rewards. South Koreans are one of the world's best educated people.

Double-check

Review

1. Why do some Koreans refer to their land as "a shrimp between two whales"?

2. What country controlled Korea for most of the first half of the 20th century?

3. What role do *jaebul* have in South Korea?

4. List three kinds of products made in South Korea that are very important to its prosperous economy.

5. What event in 1988 allowed South Korea to show off its economic miracle to the rest of the world?

Discussion

1. South Korea's main trading partners are Japan and the United States. Although still relatively small, South Korea's trade with the People's Republic of China has greatly increased in the past few years. By 1987, South Korea's trade with China was greater than North Korea's trade with China. South Korea sold refrigerators, TVs, steel, chemicals, and manufacturing technology to China. China sold coal, iron, grain, raw cotton, and silk fiber to South Korea. Do you think it wise for South Korea to continue to increase its trade with a Communist country? Support your opinion.

2. List the six keys to Japan's economic success discussed on pages 107–111. Which ones do you think also apply to South Korea? Why?

3. Compare the benefits and the disadvantages of the educational systems in South Korea and in your community.

Activities

1. The class might be divided into four teams. Over a period of a week, each team might keep a list of all the South Korean products its members have come across in their daily lives. Each team might list the products by categories such as clothing, electronics, and so forth, and make several generalizations from the information they have gathered.

2. Several students might create a bulletin-board display featuring current economic and political events in South Korea.

3. Students might form several groups, each to compose a letter to a different South Korean company. Each group's letter could ask for information about the daily life of that company's workers. Each group could report back to the class concerning the responses received.

Skills

THE FOUR TIGERS: SOME MEASURES OF STANDARD OF LIVING

Economies	Life Expectancy (years)	Literacy Rate	Income per Person
Hong Kong	76	75%	$6,910.
Singapore	73	85%	$7,410.
South Korea	69	92%	$2,370.
Taiwan	71	90%	$3,000.

SOURCES: World Bank's *World Development Report 1988*; *The 1989 Information Please Almanac*.

Study the bar graph. Then answer the following questions.

   1. Which of the Four Tigers has the highest literacy rate?

   2. The average person in which economy can expect to live the longest?

   3. What is the average income per person in South Korea?

   4. Using only these graphs, what conclusion might you reach about health care in South Korea? Why?

   5. Where did the graph makers get their statistics for this chart?

Chapter 8

Taiwan: We Will Make Anything You Want

MRS. VIVIAN YEN IS AN EXAMPLE of the economic energy of today's Taiwan. After the death of her husband a few years ago, she took over his automobile business. She knew nothing about automobiles, not even the names of the parts of the car. Still, she accepted the challenge, and now her company has developed the first Taiwanese-designed automobile. Mrs. Yen found she liked having "new challenges every day." Running a business that now has over $500 million in yearly sales brought zest to her life.

Because of the efforts of Mrs. Yen and thousands of others like her, many Taiwanese are living the good life. On the streets of the capital city, Taipei, shoppers flock to stores that stock the latest fashions and luxury goods. Other shoppers may prefer the old Chunghua Bazaar, or "haggler's alley," where buyers and sellers negotiate the prices of everything from curios to high-tech electronic products.

The government of Taiwan has encouraged businesses to invest in the microelectronics industry in Hsinchu, a city southwest of the capital city of Taipei.

This prosperity has come to 20 million people who live on an island the size of West Virginia. Its official name is the Republic of China, and its government claims to represent all of the billion Chinese who live on the mainland only 100 miles away. But Taiwan ("terraced bay") is recognized as a nation by only a few countries—making its economic success even more amazing.

Beginning again. In 1949, on the China mainland Chiang Kai-shek faced defeat in the Chinese civil war, which had been going on for over 20 years. During most of that time, Chinese Communist guerrilla forces under Mao Zedong had opposed the Nationalist party government led by Chiang Kai-shek. World War II interrupted the civil war, as both sides united to fight the Japanese. After 1945, however, China's internal struggle began again. High inflation and Chiang's corrupt officials weakened the government, and in 1948 and 1949, Communist troops overran the country.

Chiang fled with his supporters to the island of Taiwan. He set up his government at Taipei, where delegates from each of the mainland provinces met in a

Vivian W. Yen, the president of an auto company in Taiwan.

In 1965, these students marched down the main street of Taipei to honor Sun Yat-sen, the Chinese statesman and revolutionary leader who became the provisional president of China in 1911. Besides carrying the portrait of Sun Yat-sen, they also carried the portrait of Chiang Kai-shek. What role did Chiang play in the government of Taiwan?

"national assembly." Chiang declared his intention to regain the mainland. He maintained that he represented the legitimate government of China. The United States firmly backed his claim and kept Mao's government from taking China's seat in the United Nations.

Taiwan had been part of China since the 13th century. Some Chinese had moved to the island over the centuries. But it was not until the 17th century that many people from the Chinese province of Fujian settled the islands. These Chinese settlers called the land *Bao-Dao*,* "Treasure Island." After a period of Dutch colonial rule from the 1620s to the 1660s, the island was controlled by mainland China. From 1895 to 1945, Taiwan was a Japanese colony. Under harsh Japanese rule, economic development began. Roads and railroads

Farmers transplant rice sprouts farther apart to give them room to grow. What did the Nationalist government do to help improve crop production?

were built. The Japanese constructed irrigation and hydroelectric projects, coal mines, and factories. They introduced better farming methods. However, when the Nationalists arrived in 1949, Taiwan's economy still consisted mainly of rice and pineapple farming.

Eventually, two million of Chiang's supporters followed him to Taiwan. However, more than five million Taiwanese already lived there, and they resented Chiang's rule. His followers came from all of the provinces of China and spoke many Chinese dialects. They established Mandarin as the official language, while the original inhabitants spoke the dialect of Fujian province. The government was a one-party state with the Kuomintang, or Nationalist party, holding power.

In the early years, the government began a land reform program. It divided the estates of former Japanese landlords among the people. The Nationalist government set up programs to provide more irrigation and fertilizers, better seeds, and a new double-crop rotation system. Farmers have been able to more than double the amount of rice, wheat, and sweet potatoes they produce per acre. Sugarcane and tea were important cash crops. Small farms, however, still did not produce enough to support the sudden rise in population. For years, Taiwan was heavily dependent on American aid of $100 million a year.

In 1953, the government started a series of four-year plans, using American aid to establish light industries such as plastics, electrical appliances for consumers, agricultural processing, and construction materials. As the economy grew, so did the population. Like other Pacific Rim countries, Taiwan had to find a way to build an industrial base to increase its exports. To do this it needed money, and the American aid program was due to end in 1965.

Export processing zones. The Taiwanese hit on an ingenious solution to their problem. In 1966, they opened the first export processing zone (EPZ) at the southern port of Kaohsiung.* To induce foreign companies to build factories in the EPZ, the government offered them a five-year tax holiday. There were to be no taxes on imports and exports, factories, banks, or port access. As in South Korea, labor strikes were forbidden. In return, Taiwan received new jobs, foreign exchange, and technological experience. The EPZ combined the ideas of a free trade zone and an industrial park.

The experiment was a great success. Kaohsiung became one of the busiest ports in the world. "Made in Taiwan" labels appeared on Sunbeam hair dryers and Marantz stereos. Managers who were trained in the zone could then use their skills to operate Taiwanese factories outside the zone.

In addition, the zone became a magnet for farmgirls who found jobs in the factories, entering the workforce and urban life for the first time. "The traditional Chinese way of thinking that women should stay at home has been completely changed," said Kwei-Jeou Wang, a director of the development zone.

The zone provided dormitories for the women. To encourage their spirit, the factories awarded prizes for the most productive workers. Beauty contests were also part of the zone's social life, and the former farmgirls discovered such modern ways as blue jeans and lipstick. However, they also found acceptance as valued workers, and some women were promoted to managers.

The EPZ idea was so successful that more were established. Other Asian countries copied the idea. Although low-skilled clothing manufacturing predominated in the early years, electronics soon took its place. Wages rose for workers in the zones. All the while, the Taiwanese were developing their own businesses outside the zone areas.

A "hungry" people. In 1971, political disaster struck. By an overwhelming vote, the Republic of China (Taiwan) was removed from the United Nations and Mao's People's Republic of China replaced it. Taiwan became a pariah nation—the number of countries that diplomatically recognized it as the Republic of China shrank yearly. Even the United States, in 1979, recognized the People's Republic of China and cut formal diplomatic ties to Taiwan. Although the People's Republic seeks unification with Taiwan through peaceful means, Taiwan's future was placed in jeopardy.

Political humiliation only gave the Taiwanese "a

A worker in Taiwan's ship repair industry.

hungry spirit" to achieve. They responded to diplomatic defeat by increasing their efforts on the economic front. They improved the railroad system and built heavy industries such as steel plants and chemical factories as well as shipyards and nuclear power plants. Knowing they could not rely on others, they stepped up their manufacturing for export. Because their wages were no longer as low as before, they knew they had to move up the technological ladder to stay competitive.

In 1980, the government established a Science Park to attract computer companies and bring that technology to Taiwan. Red and white banners flew over Taipei urging the people to "meet the coming of the Information Era." Modeled on the earlier EPZs, the Science Park hoped to attract the best and brightest, and it did. One of the first companies to locate there was high-tech Wang Laboratories of Massachusetts, and more followed.

The quick-learning Taiwanese soon moved from assembling foreign computers to making their own. Stan Shih is one of Taiwan's success stories. He began by selling duck's eggs and watermelons on the street. He showed his creativity by inventing a pen watch. Using the profits from that, he began making Taiwan's first electronic calculator. Now he is the head of Multitech Corporation, the largest computer manufacturer in Taiwan.

In the 1980s, Taiwan moved heavily into high-tech industry. Science Park became its Silicon Valley and attracted to Taiwan some talented Chinese Americans, who were educated in the United States but felt comfortable working in China.

A roller crushes counterfeit computers in a Taiwan government effort to stop such products from entering the marketplace.

Textile executives and factory workers demonstrated outside the U.S. mission to Taiwan. They were protesting American policies that were trying to make Taiwan's products more expensive for Americans to buy.

Even so, Taiwan still produced such standards as bicycles, shoes, clothing, and home appliances. The slogan of Taipei was, "We will make anything you want."

By 1988, Taiwan was the 13th largest trading power in the world. Taiwan uses much of the money it earns from selling its goods to other countries to buy the raw materials its industries need. The textile industry depends on imported raw cotton. Indonesia sells its bauxite ore to the Taiwanese aluminum industry. Saudi Arabia provides the oil needed by Taiwan's growing industries. However, by the late 1980s Tawan had built up enormous trade surpluses. In 1987, the Taiwan government was the world's largest buyer of silver and gold bullion. Some regard precious metals as the safest form of wealth. However, Taiwan also held foreign cash reserves second only to Japan.

Taiwan's main market was the United States. Because its products were lower priced than American-made ones, Taiwanese imports flooded the United States. Americans became concerned about lost jobs and the trade deficit with Taiwan. Strains increased because Taiwan restricted the import of American goods—even after it had huge trade surpluses in its favor.

Another sore point was Taiwan's pirating of books and tapes for the world market. Taiwan's businesses duplicated and sold American products without paying for patent rights and copyrights. Taiwan was also a center for knock-offs of designer clothing—illegally manufacturing stylish clothes with foreign labels.

To appease the United States, Taiwanese went on buying missions to the United States. Taiwan also promised to change its trade practices in an attempt to preserve the once-close relationship between the two industrial countries.

A Confucian society. Confucian values are deeply embedded in Taiwan. Family loyalty is the strength of hundreds of thousands of family-owned businesses. In contrast to Japan and South Korea, small family businesses form the backbone of Taiwan's economy. A young person starting out will work for a foreign company for a few years to get experience and then open his or her own business. Traditionally, the Chinese like to be their own bosses. "Rather be head of a chicken than tail of an ox," runs an old proverb.

The fact that so many businesses are family-owned means that people are willing to work long hours, since they are working for their personal stake. Overtime is taken for granted. People are willing to work for less to succeed. The Confucian work ethic is as strong in Taiwan as in the other economic Tigers.

The pattern of many small businesses gives Taiwan's economy great flexibility. It can adjust to changes in a short time. It is not uncommon for firms to switch from making one kind of product to another in a short time. In the long run, however, smallness could pose problems. Small companies do not have enough ready money for the extensive research and development that larger corporations can afford. Nor can small companies take advantage of a large talent pool or the efficiencies of mass production.

Taiwanese share the Confucian love of education. September 28 is Teacher's Day, a national holiday. The date commemorates the birthday of Confucius, "the teacher of all generations." Taiwan's literacy rate is very high, and it has the advantage of a young, educated work force. Intense study means there is not much time for recreation and the sports that junior and senior high school students like—baseball and volleyball.

Taiwan University is noted for its high-quality physics department. However, Taiwan's educational weakness is in postgraduate studies. There are some universities that have begun to offer the highest levels of scientific training. But many ambitious and talented Taiwanese apply to graduate schools in the United States—now sometimes called "the other mainland." A street in Taipei is called Cram School Alley, because students go to training schools there to practice taking the entrance tests for American universities. Unfortunately, once its students go abroad, about three-quarters of them do not return. Attracted by the high salaries offered in the United States, they take jobs there. This "brain drain" hampers Taiwan's development of high-tech industries.

Frugality is another quality of the Taiwanese. They

have the highest rate of personal savings in the world. Fully 35 percent of earnings are saved, in contrast to about 4 percent in the United States. One reason for the high savings rate is the uncertainty of Taiwan's future. Cash can quickly be transferred to a safe place overseas should the island become part of the People's Republic of China.

But just as in Japan, the younger generation of Taiwanese may be more free-spending than their elders. Credit cards, formerly unknown there, were introduced in 1989. If Taiwan adopts the "buy now, pay later" attitude of the United States consumer, then Taiwan

The National Museum Palace contains many of the ancient art treasures taken from the mainland by the Nationalists in 1949. What cultural values have helped Taiwan succeed economically?

may become a bigger customer of American and Japanese consumer goods.

Mainland fever. Soon after the death of Chiang Kai-shek in 1975, his son Chiang Ching-kuo became leader on Taiwan. The 85 percent of Taiwanese who had been born there ("islanders") resented the restrictive government, led by those who had arrived in 1949 ("mainlanders"). Chiang the son loosened the government's political control. He encouraged more islanders to join the Kuomintang party, which formerly was controlled by mainlanders. Chiang Ching-kuo also ended the special laws put into effect in 1949, which had given the government dictatorial powers. An opposition political party, the Democratic Progressive party, was legalized. Chiang appointed as his successor Lee Teng-hui, a native Taiwanese educated as an agricultural economist at Cornell University in New York State. Lee, who had already served as mayor of Taipei, governor of Taiwan Province, and vice president, became the head of government after Chiang's death in 1988.

After decades of rule by a one-party government, many people are thrilled at the growth of democracy. Others are skeptical and think freedoms won't survive. Still others are taking a wait-and-see attitude. Antonio Chiang, an editor whose magazines were often seized by the government when it didn't like what he published, says: "If a kite is flying high it may look like a bird, but people are still pulling the strings. Some people say this is a bird, some say it is a kite. I don't know yet."

On the other hand, attitudes towards Communist rule on the mainland—the government of the People's Republic—have slowly changed. Under Chiang Kai-shek, the Kuomintang's position had been clear: No negotiation, no compromise, no contact. The People's Republic, for its part, has always said that Taiwan is part of China.

But the two governments gingerly established limited contact in November 1987. In Taiwan, this set off "mainland fever"—books and movies from the mainland were much in demand. Between 1987 and the beginning of 1989, over 400,000 people from Taiwan visited the mainland, where many still had relatives.

These visits caused mixed feelings. On the one hand, some have rediscovered their roots. On the other hand, many found the People's Republic backward and chaotic. "The more you see in the mainland, the more you treasure what you have on Taiwan," said a Kuomintang official. This caused much discussion about the future of Taiwan. Some favor an eventual peaceful reunification; others oppose it. Some native Taiwanese have long harbored a desire for full independence. Now that they have a greater say in the government, they may seek to carry out that goal. Taiwan's ultimate fate remains unsettled. Its booming economy rests on a political base of great uncertainty But the Taiwanese' s great talent for adapting to change will undoubtedly serve them well.

More About Sports in Taiwan: Shuttlecock Kicking and Baseball

Hsiang Kuo-Ching of Taiwan's Far East Team slides safely home during Little League World Series action. What other sports are also popular in Taiwan?

"Find a coin... glue a chicken feather on one side... wrap the coin in thick paper or fabric leaving the feather sticking out like a tail. Throw it up in the air and kick it when it comes down. The goal is to kick it the longest without letting it fall on the ground," said Shr-Jin Wei.* He was describing shuttlecock kicking, the ancient Chinese game he played in junior high school when he was growing up in Taiwan.

Taiwanese students play many modern games. In fact Taiwan's Little League, Senior League, and Big League baseball teams have won many world titles in the last 20 years. But most junior and senior high school students in

Taiwan also learn traditional Chinese activities, such as shuttlecock kicking, Go, and Kung Fu.

"Everybody knows how to play Go, or Chinese chess. From elementary school to college it is played," said Shr-Jin.

Go is believed to have been invented in China around 2,000 B.C. It is played with two players, each of whom have a 181-piece set of either white or black stones. Go is played on a rectangular board called a "territory." The board is divided into a grid of 361 small squares. The object of Go is for one player to capture more territory than the other by surrounding either squares on the board or the opponent's stones.

Kung Fu is a form of unarmed combat that includes physical exercise and mental and spiritual discipline. It's beginnings have been traced to a Chinese dynasty around 1000 B.C. It is taught today to Taiwanese boys in physical education classes. They learn the physical movements of Kung Fu for exercises, says Shr-Jin. The mental and spiritual parts are not taught, he said.

Basketball and volleyball are the favorite modern sports of Taiwanese boys and girls. These sports are especially popular with junior and senior high school students who live in cities. In the countryside, where there is space for playing fields, boys also play baseball, soccer, and rugby. Rugby is an English game similar to American football. Girls who live in the country play shuttlecock kicking, softball, jog, and run more than girls who go to school in the city. Badminton and Ping-Pong are popular sports for both city and country girls, said Tsun-Mei Chung,* who graduated from a private girls' high school in Taipei.

Though Taiwanese students enjoy games and sports, they don't have much time to play because they study long hours. Shr-Jin says the average student spends only three or four hours a week playing sports or games. He says that one out of 100 students is on a sports team. They play about 10 hours a week. Teens in Taiwan enjoy modern sports for recreation and international competition while they preserve their heritage by playing traditional Chinese games and sports.

Double-check

Review

1. What is an export processing zone?

2. Besides the Nationalist government in Taipei, what other government claims that Taiwan is part of its country?

3. List three heavy industries that Taiwan developed.

4. What high-tech product is now a major export of Taiwan?

5. What is the major energy resource that Taiwan imports?

Discussion

1. In 1997, the British colony of Hong Kong will become a part of the People's Republic of China. Many of Hong Kong's highly trained Chinese workers do not believe the Communist People's Republic of China's assurances about maintaining Hong Kong's capitalist system of private ownership of businesses. These people want to leave Hong Kong. The Taiwan govenment and many people in Taiwan do not want massive Chinese immigration from Hong Kong. Despite a shortage of highly trained workers, many in Taiwan claim their island is already overcrowded and that some of the immigrants might be Communists. Should Taiwan welcome these immigrants or not? Why?

2. What do you think are the three most important reasons for Taiwan's economic miracle? Support your opinions.

3. Compare two traditional values of Taiwan's people with two traditional values found in your community. How are they alike and how are they different? Which values might lead to economic success? Should economic success be the main way of judging whether a society is successful?

Activities

1. Several students might role-play a meeting of top government officials and economists in Taipei. They are discussing what the Taiwan government's policies toward the many small family-owned businesses should be. Are such businesses a hindrance to further economic development?

2. Two groups of students might create large posters illustrating Taiwan's trade patterns with pie graphs. One group might research the most recent information about which countries are major buyers of goods and services from Taiwan. They could show these major buyers in a pie graph. The second group might draw a pie graph showing countries that are major sellers of items to Taiwan.

Skills

USING AN INDEX

> Taipei, Taiwan, 143, 144–146
> Taiwan, 9–10, 13, 14, 117, 129,
>   143–161, 220–221, 236
>   Chinese government at, 60
>   conquered by Japan, 57, 76
>   land, 19
>   people of, 34
>   political fate of, 156-157
>   sports in, 154, 158–159
> tariffs, 15, 89, 107–109, 211, 212
> Tasman, Abel, 210
> technology rights, 109
>   *see also* high technology
> Thailand, 13–14, 46, 54, 62, 188,
>   194–195, 222
>   agriculture, 26
>   land, 19, 20, 21
>   natural resources, 28
>   urbanization, 27
> Tokugawa family, 70–71
> Tokyo, Japan, 13, 17, 73, 86, 92,

Use the above listings from the Index to this book to answer the following questions.

1. In what order are topics listed in an index?
 (a) by importance (b) by page numbers (c) alphabetically

2. What do the numbers after each topic stand for?
 (a) ages (b) chapter references (c) page references

3. On what page might you find a discussion of student playing sports in Taiwan?
 (a) 105 (b) 158 (c) 217

4. On how many pages in this text is the topic of Taipei dealt with?
 (a) 1 (b) 4 (c) 10

5. Does this text discuss the land in Taiwan on more pages, the same number of pages, or fewer pages than it discusses the land in Thailand?
 (a) more pages (b) the same number of pages (c) fewer pages

161

Chapter 9

Hong Kong: Living on Borrowed Time

FROM HER HOME, Freda Dan can look out on the "fragrant harbor" from which Hong Kong gets its name. Boats of every description—huge cargo ships, hydrofoils, ferries, and ramshackle fishing vessels ply one of the world's great harbors. On land, Hong Kong's distinctive skyline soars with sleek skyscrapers that house banks and insurance and trading companies. They form the basis of Hong Kong's great financial empire.

Life in Hong Kong has brought prosperity to 17-year-old Freda Dan's family. Freda's mother works in an advertising agency, and her father is a government employee. Freda likes to play the piano and sing in the choir. She also likes music videos and corresponds with

An ultra-modern Hong Kong bank building.
Hong Kong is one of the world's leading financial centers.

a U.S. pen pal. Although she dreams of becoming a model, Freda is planning a more practical career in business administration or hotel management. She hopes to go to college in Hawaii or Canada.

However, Freda's future is clouded by uncertainty. In 1997, Hong Kong will be returned to the People's Republic of China. China, the world's most populous Communist nation, will absorb Hong Kong, whose 5.6 million people enjoy one of the world's purest forms of capitalism. The contrast between the two systems is great. Under communism, the government controls the major means of production, such as factories. Under capitalism, property is owned by individuals or companies. Under capitalism, prices, production, and distribution of goods are not determined by the government. They are determined by competition in a free market.

The lion and the dragon. The British seized Hong Kong in 1841, after a short war with China. The following year, the two countries signed a treaty granting Hong Kong Island to Great Britain "forever." Hong Kong didn't seem like much of a prize. Located off the southeast coast of China, it was "a barren rock with hardly a house upon it," according to British foreign minister Lord Palmerston. A visiting journalist commented that "hawkers of every description abound"—a characterization that is still true today.

The British turned this barren rock into treasure. They developed its harbor and made it an entrepôt,* or relay port, for exchanges between China and the West. In the 19th century, ships carrying goods from the West and opium from India were unloaded there and reloaded with Chinese goods bound for Western markets. Hong Kong prospered by handling other people's products, for it had none of its own. From the beginning, it was a free port, where ship owners did not have to pay duties, or taxes, on their cargoes. But they did have to pay for storage, shipping, loading and unloading, and all the many services that a port provides.

Hong Kong became so crowded with ships that the British sought to expand its colony—and again China was too weak to resist. It was forced to give up nearby Kowloon peninsula in 1860. In 1898, China signed a 99-year lease granting the British an even larger chunk of the mainland, called the New Territories.

The three elements—Hong Kong Island, the Kowloon peninsula, and the New Territories—all form part of today's colony of Hong Kong. It covers about 400 square miles, a bit more than one-third the size of Rhode Island, the smallest U.S. state. Although 99 percent of the population is Chinese, it remains a British dependency, or colony; the head of government is a governor appointed by Great Britain. On the shield of the colony, the British lion meets the Chinese dragon.

After 1949, Hong Kong continued its entrepôt role, this time handling trade with the Communist People's Republic of China. But the Korean War chilled relations between the West and China. In 1951, the United Nations called for a trade embargo, or ban, on all goods from the People's Republic of China. This threatened more than 50 percent of Hong Kong's trade. To survive, it had to change.

Made in Hong Kong. In the 1950s and 1960s, Hong Kong faced two major problems—the sudden rise in its population caused by Chinese refugees, and the loss of its position as the entrepôt for China. To accommodate the refugees, the government allowed the construction of multistory buildings, which had previously been forbidden because of fear of fire. Chinese from Shanghai, a center of the textile trade, began to set up clothing factories in their apartments. These became known as

These garment workers are part of Hong Kong's huge clothing and textile industry. Some Hong Kong industrialists have built factories in nearby areas of the People's Republic of China where wages are only one-third of those in Hong Kong.

"flatted factories," from the British term for apartment—flat.

The flatted factories turned out textiles and clothing made of cotton, wool, and synthetic materials. They gave Hong Kong an important source of export and helped to solve its second problem. More small industries began to grow, aided by the low taxes set by the government and the absence of a minimum wage.

Hong Kong also continued to take full advantage of its magnificent deep-water harbor. Since it remained a free port, it attracted goods from many Asian countries destined for re-export. Though Hong Kong collects no duty fees, the re-export trade benefits it by creating jobs and service industries. Warehouses, trading companies, and banks are among those who benefit from the re-export trade.

Hong Kong's free trade policy, low wages, and unrestricted business also attracted industries to build factories there. Today it is one of the greatest centers of light industry in the world. It remains a major clothing center, producing everything from cheap, mass-produced shorts and T-shirts to high-fashion designer dresses. But it has branched out into other fields. Toys are a Hong Kong specialty. Today the colony is the world's largest manufacturer of toys. The colony's factories also make electronic goods, such as watches, radios, calculators, and television and computer parts. Most of the newer factories are in Kowloon and the New Territories, where there is more room than on small Hong Kong Island.

Finance and tourism. Hong Kong's freewheeling capitalism has made it an important financial center. The rich banks of Hong Kong have branches all over the world. Financial planning services have kept pace. Foreign investment, predominantly by the United States, Japan, and China, is encouraged by a lack of

One of Hong Kong's many toy manufacturers. What other light industries are important to Hong Kong's economy?

government interference and red tape. Using the latest technology, the Hong Kong Stock Exchange actively trades stocks in many of the world's businesses.

Tourism is one of Hong Kong's most important industries. Travelers on business or holiday have a wide choice of lavish hotels, where they get top-flight service. Many of the best hotels offer secretaries, translators, telex and fax machines. One hotel in Kowloon has a fleet of Rolls-Royces that take guests to and from the airport, shops, and the business section. Hong Kong is popular for international business conventions, both because of its facilities and its shops. It also has more than 30,000 restaurants. Besides Chinese food, they offer food that is Japanese, Korean, Thai, Burmese, Indonesian, Vietnamese, Filipino, Indian, or even French, Italan, and Mexican!

Hong Kong is also a shopper's paradise, both for what it makes and what it imports. Hong Kong tailors can take measurements and make a fine suit in three days or less. Shoes and boots also are made to order. The center of the island is the city of Victoria, commonly known as Central. Stores on its Hollywood Road sell goods from all over the world—Paris designer fashions, Swiss watches, and the latest Japanese electronic products—many of which cannot be bought in the United States. On Ladder Street, named for its steep steps, passageway stalls sell cut-rate odds and ends that may come from anywhere.

"May you be blessed with prosperity." The pursuit of profit is the main business of Hong Kong. From the lunchtables of the finest restaurants to the fish markets of the New Territories, the talk is about money and prices. The spirit of Hong Kong is to gain as much personal wealth as possible. People want their profits in the short term rather than the long term. This attitude has brought a standard of living higher than any other

The latest designs in high-fashion clothing are being presented at a dinner for department store merchandise buyers.

Asian economies except Japan and Singapore. But there are still extremes of wealth and poverty in Hong Kong.

In contrast to polite manners in Japan and South Korea, flaunting one's wealth is acceptable. It is not considered rude to mention how much one's home cost to build. Furs and designer clothes are worn to show off one's wealth. One Hong Kong stockbroker wears huge diamonds in place of buttons on his tuxedo. Hong Kong has the highest per capita ownership of Rolls-Royces in the world.

The Confucian work ethic is alive and well in Hong Kong. As the colony's prosperity has grown, more people have entered the middle class. Opportunities have opened up for women as well. "Sex discrimination is inevitable," says Freda Dan, "but in Hong Kong, a talented woman can get a good job. Everyone's very competitive and very hard working. So many immigrants come here from China, because they want a

The Hong Kong skyline. Between the mid-1960s and the mid-1980s, the value of Hong Kong's re-exports increased almost 14 times. Today, Hong Kong is the world's busiest port.

better life. That's what makes Hong Kong such an exciting place to live!"

Confucian ideals are found in the attitude toward education, with nine years of school required for all children. The demand to do well is deeply felt, and parents push academic achievement. Parents play a strong role in their children's lives. Although many teenagers date, Freda Dan says she doesn't because her parents are "very selective" about the boys she meets.

"One country, two systems." In the 1980s, the talk of Hong Kong was what would happen when Great Britain's 99-year lease on the New Territories ended. Great Britain and the People's Republic of China began to negotiate a new status for the colony. Because Hong Kong Island and Kowloon could not support themselves without the New Territories, Britain decided it would give them back at the same time, with certain safeguards for the people of Hong Kong.

In 1984, Britain and the People's Republic came to

an agreement. The people of Hong Kong were not asked to vote on this British-Chinese decision. However, the freedoms of speech, press, religion, travel, and assembly are to be guaranteed by law. Hong Kong will become a Special Administrative Region of the People's Republic of China with a "high degree of autonomy." Translated, this means that Hong Kong will be able to keep its capitalist system for at least 50 years, until the year 2147. China's leader, Deng Xiaopeng, called this solution "one country, two systems."

Even before the agreement, Hong Kong had carried on trade with the People's Republic. China has benefited from having a capitalist trade port on its coast. It sends almost a quarter of its exports through Hong Kong and has its own banks and about 2,000 of its companies there, tying it to the world economy. More than 2 million people in the nearby Chinese province already work directly or indirectly for Hong Kong companies. Several hundred thousand mainland Chinese visit Hong Kong each year. Many are tourists on group tours. Others are business people. On the other hand, over a million people from Hong Kong visit mainland China each year. Prices are much lower in China, so poor people make weekend trips by train for shopping and for doctor and dentist appointments. In 1989 many people in Hong Kong protested against the Chinese government. It had killed some of the students and workers demonstrating for more democracy on the mainland.

Freda Dan is preparing for the day when Hong Kong will be part of China. She is now studying Chinese Mandarin in school. "Hong Kong is the gateway to China," she says. "The Chinese government has promised that we will continue as the go-between in China's relations with the West." Freda has never been to the mainland. But she is optimistic that Hong Kong can live with and be part of the People's Republic of China.

Double-check

Review

1. At present, Hong Kong is a dependency of what country?
2. What is an entrepôt?
3. List three important kinds of products manufactured in Hong Kong.
4. What is meant by the phrase "one country, two systems"?
5. Why is the year 1997 important for Hong Kong?

Discussion

1. The tourist industry and the financial services industry are both very important to Hong Kong. List some of the kinds of businesses connected with each of these industries and discuss which industry might be more important for Hong Kong's economic future. Support your opinion.

2. Do you think that the people of Hong Kong should have had the right to vote on the agreement signed between Britain and the People's Republic of China in 1984? Why or why not? How do you think they would have voted? Why?

3. What similarities are there between the economies of Hong Kong and the United States or Canada? What differences?

4. The author Han Suyin has described Hong Kong as "living on borrowed time." What do you think she meant by this phrase?

Activities

1. Several students might chart, for a week or two, the rise and fall of stock prices of particular U.S. and Hong Kong stocks (or the stock price averages as measured by the Dow Jones Industrial average in the United States and the Hang Seng average in Hong Kong). The students might present the chart to the class and make some hypotheses about the relationship between the prices of stocks on the two stock exchanges.

2. Several students might use atlases and encyclopedias to draw a large map of Hong Kong for the bulletin board. Be sure to show Kowloon, the New Territories, and Hong Kong Island as well as their position in relation to the People's Republic of China.

Skills

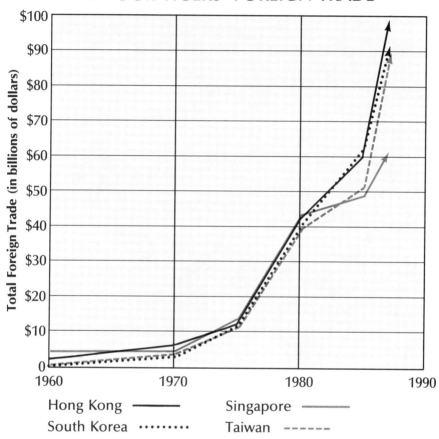

SOURCE: U.S. CIA *Handbook of Economic Statistics, 1988.*

Study the graph. Then answer the following questions.

   1. Which of the Four Tigers was the leading foreign trader in 1960?

   2. By 1970, which country had taken the lead in foreign trade?

   3. By 1987, which country had the least amount of foreign trade?

   4. Based on what you have learned in this chapter, how would you explain Hong Kong's foreign trade trends over the past 20 years?

   5. If each of the Four Tigers continued the same trade trends they had established throughout the 1980s, which country do you think might be the leading foreign trader by 1999?

Chapter 10

Singapore:
A Mighty Ministate

SINGAPORE'S LEADERS are pinning their hopes on students such as 17-year-old Victor Chua. Their goal is to make their tiny island nation of 2.7 million people the "brain center" of Asia. Singapore, less than one-half the area of Los Angeles, lies at the southern tip of the Malay Peninsula and has historically been a center of Asian trade.

Victor's ethnic background is Chinese, like that of about three out of every four of Singapore's people. He actually speaks German better than Chinese because his family lived in West Germany for six years. His father was posted there as a member of Singapore's Economic Development Board. Victor's mother is a teacher.

Recently, the prime minister's son visited Victor's school. He told the students that education was crucial for Singapore's development and its drive to be the center of Asian business. Victor is studying hard to do his part.

A statue of Sir Thomas Stamford Raffles, who founded the port of Singapore on the site of a small fishing village in 1819.

City of lions. In 1819, Sir Thomas Stamford Raffles of the British East India Company was looking for a place to set up a trading post. He found Singapore, or "the City of Lions," with one of the best harbors in the region. He negotiated a deal with the Malay *temenggong** who ruled the region. Singapore became a free port and strategic outpost of the British Empire.

The colony thrived and its population grew; Chinese and Indians now mixed with the original Malays. To the Chinese, Singapore was part of Namyang, the Chinese name for "the lands of the southern seas," or Southeast Asia. For the last three hundred years, Chinese have emigrated to the countries of Southeast Asia. They have played an important role in trade and business.

One of the leaders opposing British rule after World War II was Lee Kuan Yew, a Singaporean of Chinese descent. Lee united Communists, traditionalists, and labor leaders into the People's Action party. In 1959, Singapore was granted limited independence, and Lee became prime minister. Four years later, Lee cut all political ties with Britain and led Singapore into the four-nation Federation of Malaysia—which you will learn about in Chapter 11.

A majority of the people of the Federation of Malaysia were of Malay descent. There were cultural conflicts between Malays and the Chinese minority population. Singapore was the only nation with a majority of Chinese, and tensions grew between it and the other nations. In 1965, Singapore was asked to leave the Malaysian Federation. In a tearful speech, Lee rallied his people to try to survive as a tiny state with no natural resources in a hostile region.

The busiest port in Southeast Asia. Singapore had only two resources—its harbor and its people. The grand and spacious harbor at the entrance of the Strait of Malacca lies at a crucial point between the Indian Ocean and the South China Sea. Ships taking that route between the West and East Asia have long stopped there. For its people, hard work was a way of life. Lee took advantage of both resources.

He proved to be a shrewd and ruthless leader, turning on his former partners. He threw Communists in jail. He broke the power of the labor leaders and reduced restrictions on businesses' treatment of workers. Then he appealed to the workers' patriotism by asking them to work harder for Singapore's future.

Lee moved to bring in multinational corporations to boost the economy. He offered them cheap labor and government tax breaks. Less stringent banking regulations attracted international banks to open branches there.

Soon Singapore became one of the world's busiest ports. Tankers filled with Middle Eastern oil stopped there to have the oil refined into gasoline and other oil products. Singapore's dockworkers handled Malaysian rubber and tin, Indonesian coffee, and spices from all over the western Pacific Rim.

Lee's government has promoted the most efficient shipping methods to make its port one of the best in Asia. Goods flow in there to be reshipped all over the world. Singapore pioneered in the new methods of container shipping, building dock terminals designed just for that purpose. Today Singapore harbor is filled with oil supertankers, freighters, and huge containerized cargo ships.

Wages were regulated by the government, and there were no strikes. But as Singapore prospered, Lee made sure his people shared the wealth—if they worked hard. Under a program called "Rewarding the Plus Performance," high achievers were given raises of 10.5 percent; workers who were merely adequate got a 7 percent raise.

The government built high-rise apartment complexes complete with parks, shops, and schools. Four out of every five Singaporeans now live in these publicly built concrete high-rise buildings. To finance this public housing, the government established the Central Provident Fund (CPF). Employers and workers contributed almost 40 percent of their wages to the fund. In return, workers can borrow from it to buy apartments or to invest in government-approved industries. What is left is given back to the workers when they retire.

Because of its tiny area, Singapore must import all its food. As in the other land-poor Pacific Rim nations, it compensated by developing an export trade. And, as elsewhere, foreign companies financed the early development, but soon Singaporean-owned businesses were producing their own exports, including textiles, airplane parts, and computer hardware. Besides shipbuilding, processing lumber and rubber, Singapore's industries refine oil and make oil-based products.

Lee's program has worked. Within 20 years, a city that had been a colonial slum became one of the most modern in the world. In downtown Singapore rise the high skyscrapers of the financial district, called the Golden Shoe. The workers' earnings have risen so much that Singapore now has the highest standard of living in Southeast Asia—and second only to Japan's on the Pacific Rim.

Singapore, with the tallest hotel in Asia, also welcomes more tourists each year than the entire population of the country. Singapore vies with Hong Kong for the title of greatest shopping city in Asia. Tourists can shop in duty-free stores, including the futuristic shopping mall, Lucky Plaza. Some have called Singapore "instant Asia," for tourists can sample there the wares and cuisines of almost any Asian country from India to Japan.

Singapore's modern container shipping helps it compete as one of the world's major ports.

We are Singaporeans. The motto at the independence of Singapore was, "A Nation for All. Together. Excellence for All." Singapore is proud of its multiethnic character. Slightly more than 75 percent of the population is Chinese, 15 percent Malay, and 7 percent Indian. A small number of Europeans also live there.

Reflecting its background, Singapore has four official languages. All schoolchildren learn English because of its importance as the international language of trade. In the home, Chinese, Malay, or Tamil is spoken. Tamil is the language of Sri Lanka, the island nation south of India where most of Singapore's Indian immigrants came from. Taxi drivers in Singapore are often proficient in all four languages.

Singapore's multi-ethnic character is shown by the variety of fast foods sold on the street, and even more by its many different holidays. Although businesses use the Western calendar and take off such holidays as Christmas, the Eastern holidays bring out the pageantry of the city. At the Festival of Lights, Hindus decorate their houses with lanterns. At the end of the Muslim fasting month of Ramadan, Muslim families have a feast. In Singapore it is traditional for them to invite non-Muslim friends. The biggest holiday is the Chinese New Year when firecrackers and dancers dressed in dragon costumes fill the streets.

At the Hungry Ghosts festival, Chinese appease the wandering ghosts by prayers and burning paper money. At Chinese funerals it is traditional to leave paper food so the dead will have something to eat in the afterlife. Today, Singapore funerary craftspeople do a brisk trade in turning out paper models of expensive European cars, yachts, and mansions—so that the dead can enjoy as high a "standard of living" as they enjoyed in life.

About 4 percent of the people of Singapore are Hindus. Here some of them worship at the Sri Mariamman Hindu temple. About 27 percent of the people of Singapore are Buddhists. Members of another large religious group, the Muslims, make up about 16 percent of the population.

Excellence for all. As the wages of Singaporeans rose, Prime Minister Lee emphasized that now that the country's work force was not the cheapest, it had to be the smartest. "Excellence encapsules in one word how Singapore can survive in a very competitive world," said Lee at the National Day Dinner in 1987.

This has put great pressure on school children. The schoolday officially runs from 7:45 A.M. until 1:15 P.M. But almost everyone stays later in the afternoon for extra classes or extracurricular activities. Students specialize in one of three areas—science, commerce, or art. Classes are divided between large lectures and small groups of 25 or fewer.

Students are serious about their work. Victor Chua has chosen to specialize in science at the National Union College, one of Singapore's more prestigious senior high schools. Most students take three subjects, but Victor is taking five—English, math, biology, physics, and chemistry. He hopes to go to Cambridge University in England after graduation and pursue a career in science.

Although Victor received a perfect score in the examination to enter his school, he still feels pressure to study. "If you look around," he says of his classmates, "everyone's about to fall asleep. That's because most of us have been up all night studying. Exams are only 14 weeks away and the competition is very tough." He doesn't remember when he had the free time or desire to watch television.

Singapore's future. Prime Minister Lee has been the only ruler an independent Singapore has ever known. Although he favors a free market, he does not follow the same policy in governing. His influence is felt in all walks of life. Continual government campaigns try to change or influence Singaporeans' way of life.

Wall posters are everywhere, promoting the govern-

A worker in the Singapore semiconductor microchip industry. Only the United States, Japan and Malaysia export more microelectronics.

ment's latest campaign. One announces, "Keeping our city clean is a national objective." Indeed, the city is one of the cleanest in the world—partly because stiff fines as high as $250 penalize anyone who discards a cigarette or other litter. The "PM" does not like hippie life-styles, and for a time posters declared, "Males with long hair will be served last." Another poster reminds people, "Make courtesy our way of life." The government sponsors family planning, with the slogan being "Two Is Enough."

Although Lee's government has brought undreamed of prosperity to Singapore, some are tiring of the constant campaigns. Lee tolerates no criticism. Newspapers are censored if they offend the prime minister. Even so, his popularity has grown somewhat weaker. In 1989, when he was sworn in for another term as prime minister, Lee told the country that he would not run again.

Singapore's success has shown what will and determination can do in the face of great odds. Hard work and determination have built an economic mini-power in 20 years. Singapore has shown what can be accomplished even by a state with virtually no natural resources. And, just as Japan inspired the Four Tigers, the success of Singapore, and the rest of the Four Tigers, is stimulating growth in other parts of Asia as they, too, seek the good life.

More About Asian Names

Lee Kuan Yew is the undisputed ruler of Singapore. Takeshita Noboru was the prime minister of Japan. Can you guess which names are their *surnames*? (A surname is the name held in common by members of a family.) If you guessed the names Lee and Takeshita, you are correct. Chinese and Japanese family names are written and spoken before their *given names*—which in many non-Asian countries are the first and middle names. This tradition is the opposite of most non-Asian countries. In most non-Asian countries, surnames are written and spoken after the other names.

Asian surnames can refer to the place the family comes from or to its occupation. But most do not. Asian family names are usually not related to a family characteristic. The name Lee originally meant "pear tree" or "plums" in Chinese. The name Chang has four meanings in Chinese and a different meaning in Korean. Wang, a common surname, can be traced to an ancester who was a king. The Chinese characters are also used by the Japanese and Koreans. The name Take means "bamboo tree/stem." The name Suzuki means "bell/trees." Many Japanese last names include the word for "rice field." The name Fukuda means "good fortune/rice field."

In Malaysia and Indonesia, the surname often comes last. However, first names are frequently used, even in business.

Though most customs involving Chinese surnames are different from those in non-Asian countries, there is one Chinese custom that American women have begun to adopt. Chinese and Korean women keep their maiden names when they marry. Japanese women take their husband's surname, as most women in other parts of the world do. But Japanese law *requires* married couples to use the same surname. Many Japanese women would prefer to keep their own family's surname and are working to change this law.

Your surname, whether it comes before or after your other names, can tell an interesting story about your ancestors and their culture.

Double-check

Review

1. From the early 19th century up until World War II, and then during the postwar period until 1963, Singapore was ruled by what European power?

2. What are the three main ethnic groups living in Singapore?

3. List three industries important to present-day Singapore.

4. What did the government do to provide housing for most Singaporeans?

5. Who has been the ruler of Singapore for more than a quarter of a century?

Discussion

1. What might be considered the three most important reasons for Singapore's economic success? Support your opinion.

2. In the mid-1980s, Singapore declared that it did not have to allow a free and unrestricted press. The government began to prevent the wide distribution of foreign newspapers and magazines that did not print the full text of Singapore's official replies to criticisms. By 1988, a Singapore court supported these government actions. The prime minister stated that he could not allow American publications, which were used to criticizing their own American government, to be so critical of the Singapore government. Should any government be allowed to control the freedom of the press of foreign publications distributed within its borders? Support your opinion.

3. Should Singapore serve as a model of economic development for other less developed countries? Why or why not?

Activities

1. A committee of students might collect stories and pictures about life in Singapore today from newspapers and magazines and post these on the bulletin board for discussion.

2. Some students might role-play a presentation given by Sir Thomas Raffles and Lee Kuan Yew as they set forth their ideas and plans for Singapore to a committee that will question them on their policies.

3. Some students might research and report to the rest of the class about several of the following industries in Singapore: shipbuilding, oil refining, electronics, banking, textiles, lumber processing, tourism, rubber processing.

Skills

EXPORTS OF MICROELECTRONICS PRODUCTS

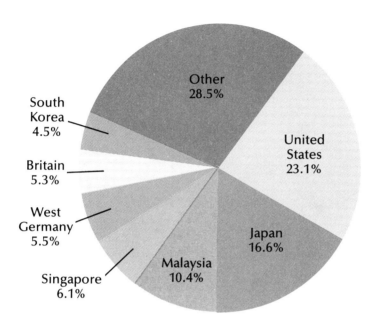

NOTE: Microelectronics products are goods with tiny electronic circuits and parts.

SOURCE: U.S. CIA *Handbook of Economic Statistics, 1988.*

Study the graph. Then answer the following questions.

 1. How is the term *microelectronics* defined in this graph?

 2. Which country exports more microelectronics products than any other?

 3. How many of the world's countries export more microelectronics products than Singapore does?

 4. How many of the top seven exporters of microelectronics are located in the western Pacific Rim area?

 5. What percentage of the world's total exports of microelectronic products does Singapore export?

4
ON THE SHORES OF THE PACIFIC

Chapter 11

The Little Tigers: Natural Wealth, Growing Industries

AT THE BEGINNING OF MARCH 1989, the sultans, or hereditary rulers, of the Malaysian states met to elect a king. They chose Azlan Muhibuddin Shah, who would serve for five years. Even though he was elected, the king is still a sacred figure to some of Malaysia's people. Many believe in *daulat*,* or the supernatural power of royalty. People must not speak to the ruler directly—only to his feet. Traditionally, royalty dined on tiny morsels of food and were bathed by gentle showers like "a petal in the rain."

But today's new king is a lawyer who has already served as a chief justice in his country's highest court. Malaysia's king rules through a prime minister. Modern Malaysia's government, like the country, is a mix of the traditional and the modern.

Malaysia is the wealthiest of a group of Pacific Rim nations sometimes called the "Little Tigers." Besides

A street scene in Bangkok, Thailand. Even traffic jams in Bangkok are an example of the old and the new.

Malaysia, they are Indonesia, Thailand, and the Philippines. The Little Tigers want to imitate the success of Japan and the Four Tigers. And the Little Tigers have something that their wealthier neighbors do not—important natural resources.

These four countries, plus Singapore and Brunei, are part of the Association of Southeast Asian Nations (ASEAN). Its purpose is to promote economic and political cooperation between the member nations.

Malaysia: Rich in natural resources. Malaysia, a nation of 17 million people the size of New Mexico, supplies 40 percent of the world's rubber, 30 percent of its tin, and 60 percent of its palm oil. Some 70 percent of the land is covered by tropical rain forests. The South China Sea separates the two largest land areas of Malaysia. West Malaysia is the southern part of the Malay Peninsula. East Malaysia includes the two states on the island of Borneo—Sarawak and Sabah.

In 1957, Malaya became independent of Britain, its colonial ruler. Six years later, Malaya and its neighbors Sarawak, Sabah, and Singapore joined together to form the Malaysian Federation.

The Federation was an uneasy coalition because many Malays did not like the economic dominance of the Chinese. In 1965, Singapore left the Federation. Malaysia, as the new nation now became known, took its present shape.

Sons of the soil. The Malays form the largest ethnic group in Malaysia. Along with other indigenous people, they compose almost 60 percent of the population. They are called *bumiputera** or "sons of the soil." Their language, called Bahasa Malaysia, is similar to that spoken in neighboring Indonesia. They are primarily farmers and fishing people and they practice Islam.

The other two population groups are the Chinese

A fishing village in Malaysia.

(about one-third) and Indians (about 10 percent). They practice Buddhism, Hinduism, and many other religions. The Chinese form the business and trading class of the country, and the Indians are prominent in such professions as law and medicine.

Kampongs and cities. Less than 40 percent of Malaysia is urbanized. The vast majority of the bumiputeras live in *kampongs*,* or villages. They are farmers, fishing people, or laborers. They tend to stick to the ways of tradition. Their lives are rich in religious and folk rituals. Most men and women still wear the sarang, a loose ankle-length shirt. In the countryside, the government has set up crafts centers to make sure the traditional practices of woodcarving and weaving are not forgotten. The large cities contain the majority of the Chinese and Indian population. The capital, Kuala Lumpur (called simply "KL"), is the political, social, and economic heart of Malaysia.

Growing industries. The government is striving to increase the country's manufacturing. In the 1960s, Penang, an island off the peninsula's west coast, became a free trade port. It was used to attract industry and trade to the country. It was the start of a program that sought to increase the industry of this resource-rich country. Malays were sent abroad for education in business. Joint ventures, particularly with Japan, were launched. Today, Malaysian factories produce electronic parts and products. With Mitsubishi, Malaysian companies developed an automobile that they hope to introduce in the export market.

Malaysia has progressed so that the average per capita income is over $1,000. This increase has allowed people to spend more. Kishu Jethanand, a store owner in KL, says, "Business is booming. The cake is now bigger so everyone can have a reasonable slice."

Brunei. A tiny neighbor, the Islamic nation of Brunei, has commercial and cultural links to Malaysia. Brunei gained its independence from Britain in 1984. On the northern coast of Borneo, nearly surrounded by Sarawak, it has only 200,000 citizens. But, thanks to massive oil and natural gas resources, it is one of the world's wealthiest nations. Sir Muda Hassanal Bolkiah is the Sultan of Brunei. The young sultan, whose family has ruled for five centuries, lives in a $300 million palace and is one of the richest men in the world.

Indonesia: Struggling to develop. "Land clearing is quite an expensive investment," says Mohammed Hasan. It's been 15 years since he started his first cacao plantation in Indonesia. The cacao seed is the source of cocoa and chocolate. Hasan's investment of both money and hard work has paid off. Today he has concessions of forest land and old rubber plantations in many parts of Indonesia.

Hasan's firm, Hasfarm Products, has been experimenting with seeds since 1979 to increase the yield of

An Indonesian mother takes her two youngest children to school. She is a dancer who lives on the island of Bali, known for its music, dance, and crafts. Bali's culture is a mix of Hindu religion and Malay culture.

the cacao crop. Now he has developed a seed that is 30 percent more productive than the standard type. Hassan supplies state plantations and local farmers with the valuable seed, as well as his own plantations. His goal is to push Indonesia's cacao industry to the forefront of world trade in the product.

Indonesia is depending on people such as Mohammed Hasan. This resource-rich nation is struggling to develop its economy and build a better life for its citizens. With approximately 180 million people, Indonesia is the fifth most populous nation in the world. Yet its per capita income is only $560, the lowest of the "Little Tigers."

Indonesia's 13,677 islands straddle the Equator for 3,300 miles—from west of Malaysia to Australia. It is the world's largest archipelago.

More than 60 percent of the population, over 100 million people, live on Java, an island the size of Alabama. This makes Java one of the most densely populated areas on earth. Other parts of Indonesia include Sumatra, the largest island; the southern two-thirds of Borneo, called Kalimantan; and West Irian, on the western half of New Guinea.

Unity in diversity. "Unity in Diversity" is an appropriate motto for Indonesia, where over 200 cultural groups share the archipelago. The national language, Bahasa Indonesia, is Malay mixed with snippets from the many Indonesian languages and dialects. It is taught in schools and used in government and trade, but most Indonesians speak their own dialect at home.

Indonesia has a rich cultural heritage. A belief that spirits occupy all living things and objects is part of Malay tradition. Both Hinduism and Buddhism were brought to the islands by Indian traders. Many of Java's shadow plays, or *wayang*,* celebrate the adventures of a Hindu god, Rama. Java is also the site of Borobudur, the largest Buddhist monument in the world. Built in the 9th century, the walls of its nine rock terraces are carved with scenes from Buddha's life. Each year on Buddha's birthday, monks from other Asian countries come there in pilgrimage.

Islamic traders brought their religion in the 15th century. Today about 90 percent of Indonesians practice Islam. But Indonesian Islam is mixed with the earlier religious beliefs.

Indonesia won its independence in 1949 after a bitter struggle with the Dutch, who had ruled the islands they called the Dutch East Indies.

The new order. In 1968, Indonesia began a series of Five Year Plans designed to reverse Indonesia's economic decline. President Suharto relied on a group of Indonesian economists who had studied at the University of California at Berkeley. The planners turned from economic nationalism to encouragement of foreign investment. Although the state maintained some monopolies, there was a greater emphasis on spurring the private economy. The army enforced political stability to allow the country to develop.

Suharto's primary goal was to enable his country to feed itself. Because of its massive population, Indonesia was the world's largest importer of rice, the chief food. Yet Indonesia has rich soil—the gift of its many volcanoes. Volcanic lava has formed nutrient-rich topsoil on the land. Also, the country's year-round 80-degree temperature makes it possible to grow two and even three crops a year.

The government provided its rice farmers—traditionally women—with new varieties of rice. The women set out the plants in groups of three, tying them together in the wet *sawah*,* or rice paddy. The new rice brought higher yields, though it required more fertilizer. At the same time, the government built roads, bridges, and markets in the countryside. In 1985, for the first time, the country produced enough rice to feed itself.

The most ambitious government project was the construction of Palapa, a satellite telecommunications system that cost over $1 billion. It linked Indonesia's countless islands in a countrywide television network. The government gave a television to every village, so that the community can gather to watch on the set in the headman's home. Antennas poke up from huts in rural areas where roads do not yet exist.

Efradus, a 16-year-old boy in a village in the jungles of Irian Java, said, "Eight years ago we didn't know what country we were in. We didn't know Irian Java, electricity, movies, TV, or money." Now they know all these things and that they are Indonesians.

Suharto says Indonesia must become less dependent on its oil wealth. The country produces rubber, tin, tea, coffee, palm oil, and other products. It also has the world's second-largest expanse of rain forests, which supply much of the world's timber. Plans are under way to turn these non-oil resources into manufactured goods.

A park in Bangkok, captial of Thailand.

Thailand: A booming nation. "I am a modern man and I like something up to date," says Petch Osathanukroh. Petch is the successful chairman of his own company, Spa Advertising, in Bangkok, the capital of Thailand. Petch's hobby is writing music. He has built a completely equipped music studio in his home to make cassettes of his songs. Petch was educated in the United States, where he acquired his taste for hamburgers and designer clothes, both of which he can now get in Bangkok.

But Petch not only enjoys the good life himself. Recently he treated his 120 employees to a yacht trip to the Thai city of Phuket. "If the company keeps on doing well, it will bring more and more benefits to the staff," he says.

Petch and his employees are enjoying the good life because of the economic boom in Thailand—literally the "land of the free." Thailand is the only Southeast Asian nation never controlled by a Western nation. Thailand's 55 million people are aiming to move up from developing-nation status to become a newly industrialized economy (NIE). With per capita income reaching toward the $1,000 mark, Thailand still has a way to go, but it is moving up quickly.

A nation of farmers. In most developing countries, farmers are barely able to put food on their tables. In Thailand, the farmers produce enough to put food on other people's tables, too. Thailand's farm families make up two-thirds of the nation's population. Most work tiny plots that average just seven acres in size. Yet their crops account for 60 percent of the nation's exports.

This Texas-sized country is the world's number one exporter of rice, pineapple, and cassava (a plant used for making tapioca and bread). It's also a major supplier of rubber, teakwood, cotton, corn, and other agricultural products. To these products is added the opium poppy, grown on Thailand's northern border with Burma and Laos. That region, known as the Golden Triangle, produces the raw material for heroin, which is smuggled to the West through Bangkok and Hong Kong.

Chasing the Four Tigers. The buying power of Thai farmers has created a growing demand for manufactured goods. And because the government takes a hands-off approach, Thailand is fertile ground for foreign investors. Thailand welcomes foreign development of manufacturing, mining, agriculture, and fisheries. The nation's abundant tin, gems, and rubber make such investment attractive. The government has protected foreign investors with guarantees of good conditions. It gives preference to projects that will provide jobs for Thais and also to businesses that will locate outside of Bangkok.

Foreign investment allowed the country to build food processing plants. Japanese and other foreign companies built plants in Thailand to assemble products ranging from cars to computers. This economic activity built on itself, creating a boom that has increased the wealth of most Thais.

The Philippines: People power to the rescue.
Many changes have occurred under the Philippines' new president, Corazon Aquino. The changes began in February of 1986, when Aquino led a dramatic, nonviolent revolution that ended 20 years of corrupt rule by Ferdinand Marcos. She was voted into office by what was called "people power." But Aquino still faces the challenges of restoring democracy and rebuilding her troubled nation's economy.

A special relationship. Aquino's success—or failure—will be closely watched by the United States. For the Philippines, a nation of 7,100 islands, is a key link in U.S. defense plans for the Pacific. America's biggest foreign military bases, Subic Bay Naval Base and Clark Air Force Base, are in the Philippines. The two nations are also linked in a special relationship, for the Philippines were once an American colony. Their relationship began as a result of the Spanish-American War in 1898. That war marked the end of Spain's 330-year hold on the Philippines and the beginning of American rule. But it was not until July 4, 1946, that the Philippines received full independence.

Picking up the pieces. Corazon Aquino, elected president in 1986, faced a society where almost 70 percent of the people lived in desperate poverty. In Manila, the capital, half the work force was without full-time jobs. In the countryside, conditions were even worse—four out of five people suffered from malnutrition.

The new Aquino government took action. A program to redistribute some of the land on large estates increased rice production and eased the hunger in the countryside. Increased aid from the United States helped to get private businesses going again. The new government encouraged joint ventures with foreign

Presdent Corazon Aquino of the Philippines. What are some of the economic changes that have occurred during her presidency?

companies to attract manufacturing plants. As confidence in the new government grew, many companies accepted the offer, and employment increased.

The island's luxury hotels for tourists stood empty during the later Marcos years. Now they welcome a greater number of visitors than ever. Aquino herself has also pushed the handicrafts industry to export more goods.

Since Aquino became president, foreign investment has risen dramatically—it more than tripled from 1987 to 1988. Much of this investment went into such crucial areas as electronics, petrochemicals, textiles, and fish farming. Chinese investors from Taiwan have fueled the construction boom. Only the future will tell whether the people of the Philippines are to enjoy the Pacific Rim's prosperity.

More About Islam

What did you do before school this morning? Chances are that you were awakened an hour or two after daybreak by the insistent buzz of your alarm clock. But if you lived in Indonesia, Malaysia, or Brunei in the Pacific Rim, your day might have begun in a very different fashion.

At the first rays of sunrise, you would have heard the haunting cry of a *muezzin*,* who announces the time for prayers and summons people to prayer.

For more than 860 million people—and perhaps as many as a billion—sunrise prayer is a ritual that occurs every day. These people, known as Muslims, practice a religion called Islam.

Islam was founded by a 7th-century Arabian trader named Muhammad. In the 13th century, commerce won converts throughout Indonesia, Malaysia, and India.

From the beginning, Muhammad and his fellow Muslims followed a set of practices that are set forth in the Koran, the Muslim holy book. The five most important of these duties are known as the Five Pillars: 1. Muslims must profess their faith in the words of the Koran: "There is no God but God (Allah), and Muhammad is the messenger of God." 2. Muslims pray five times a day, facing toward Mecca. 3. They must give to charity, called *zakat*. 4. They must fast from daybreak until sunset during the month of Ramadan, when the Koran was written. 5. Muslims should make at least one pilgrimage, or journey, to Mecca in their lifetimes.

While Muslims of all cultures practice the Five Pillars, Islam in Indonesia and other Pacific Rim countries is blended with Hinduism and Buddhism. To this day, Muhammad's tenets have not replaced village law in Indonesia. Mosques, or Islamic houses of worship, are sometimes thatched huts instead of domed palaces as in the Middle East. In Kuala Lampur, Malaysia's 1965 National Mosque has a star-shaped dome and a single minaret that rises about 240 feet from the center of a fountain area. Over 8,000 worshipers attend its grand hall at one time or another.

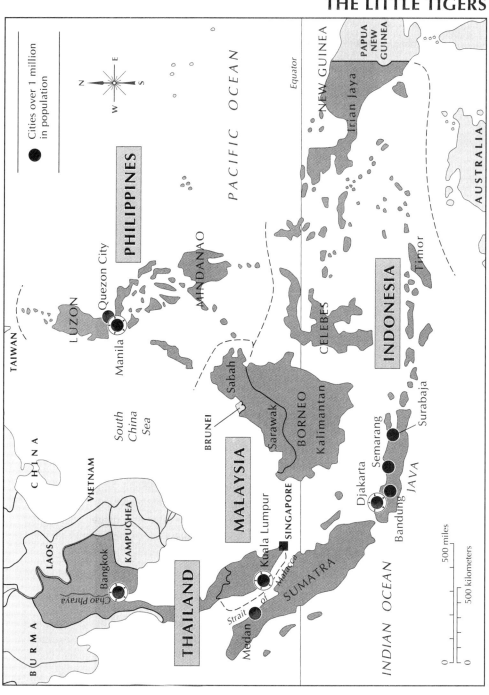

Double-check

Review

1. List the four countries often referred to as the Little Tigers.

2. Who are the *bumiputera* of Malaysia?

3. Define the term *kampong*.

4. What are three of Indonesia's major resources?

5. What country ruled the Philippines from 1898 to 1946, except for several years during World War II?

Discussion

1. Using the information in this chapter, predict which of the four Little Tigers will lead the others in economic development over the next four years. Support your opinion.

2. What role has geography played in the economic development of each of the four Little Tigers? Use the information in this chapter and Chapter 1 to provide specific examples.

3. Why is it important for developing countries to diversify their economies?

Activities

1. Several students might research the roles played by investments from Japan, Taiwan, South Korea, and Hong Kong in the development of the economies of the Little Tigers. Students might use the *Reader's Guide to Periodical Literature* to locate recent magazine articles dealing with this topic. Students might then report their findings to the class.

2. Students might form four committees, each to research life in one of the Little Tigers's capital cities: Kuala Lumpur, Djakarta, Bangkok, and Manila. Each committee might prepare a bulletin-board display with drawing, photos, and brief descriptions, which would give others a picture of the variety of life-styles found in the capital city they are assigned.

3. Several students might play the role of economists who are attending a panel discussion on economic development. The panel will compare the economic development of the four Little Tigers with Japan and the Four Tigers by using the six keys to Japan's economic development listed on pages 107—111 as a basis of the discussion. After each panel member has presented his or her comparisons, members of the class will be free to participate in the discussion.

Skills

THE LITTLE TIGERS: SCHOOL ATTENDANCE

Countries	Percentage of Secondary School Children Attending School	
	1965	1985
Indonesia	12%	39%
Malaysia	28%	53%
Philippines	41%	65%
Thailand	14%	30%

NOTE: Secondary school children for purposes of this table are ages 12 to 17 years.
SOURCE: World Bank's *World Development Report 1988.*

Study the table. Then answer the following questions.

1. For the purposes of this table, secondary school children are defined as those being how old?
    a. 10- to 18-years old   b. 12- to 17-years old   c. 14- to 18-years old.

2. Which country had the lowest percentage of its children attending secondary school in 1965?
    a. Indonesia   b. Malaysia   c. Thailand

3. Which country had the highest percentage of its children attending secondary school in 1965?
    a. Malaysia   b. Philippines   c. Thailand

4. Which country had the highest percentage of its children attending secondary school in 1985?
    a. Indonesia   b. Philippines   c. Thailand

5. In 1985, Japan had 96 percent and South Korea had 94 percent of their children in secondary school. What generalization can you make about economic development using these facts and the information in this table?

Chapter 12

Australia and New Zealand: Diggers and Kiwis

IMAGINE A SCHOOL WHERE THE CLASSROOM is twice the size of France. That's the case for Ann Scherer's 13 students, who live at isolated cattle stations (ranches) and mining camps in the Australian outback, or countryside. Although she rarely sees her students, Mrs. Scherer gives them a thirty-minute daily lesson by two-way radio from a broadcasting station in Alice Springs. "Good morning! It's music time with Miss Scherer. How do you read me?" she begins through the microphone. The students check in with a "loud and clear." Her classroom is a part of The School of the Air.

The state-run school mails each student a study program that is sent back for grading at regular periods. Books, pencils, and paper, plus the large two-way radio, are free. Parents supervise their children's study at home. The curriculum is similar to that in the United States.

Sydney, Australia. Sydney's famed opera house is located on the shore of Sydney Harbor.

New Zealand has a similar system for children in the distant regions called wop-wops. As in Australia, the school year starts in February rather than September. For south of the Equator, the seasons are reversed from those in the Northern Hemisphere.

Australia and New Zealand are in some ways similar. Both are Pacific nations whose people and culture are predominantly Western. Once tied to Great Britain, in recent years they have moved closer to other countries in the Pacific Rim. But they have followed different paths.

Convicts and sheep stations. Australia is the only nation that is also a continent. It is the smallest, flattest, and driest continent in the world. The highest mountain is only 7,310 feet high. The climate is temperate, with the south being colder than the north. And the rainfall varies from a meager 5.9 inches a year to more than 79 inches in part of the tropical north. Australia is slightly smaller that the United States, and with only 16 million people, it is very sparsely settled.

The first inhabitants of Australia were the Aborigines who migrated from Southeast Asia some time during the last Ice Age. During that time, the sea level was lower and the islands were much closer to the mainland than today. The Aborigines may be the first known sea navigators.

Aborigine art is also one of humankind's first—some of it is older than the cave paintings of Europe. The Aborigines developed in isolation from other peoples and kept their methods of hunting and food gathering. There are today 160,000 Aborigines in Australia.

Their isolation was broken in the 16th century as Dutch seafarers explored the coast. But it was the arrival of Captain James Cook in 1770 that brought a change to the Aboriginal way of life. After Cook's voyages along

Australian tourists greet a wild dolphin in the Shark Bay area of Australia's west coast. Australians are known worldwide for their love of sports and outdoor recreation.

the eastern and northern coasts, the British claimed the continent.

In 1788, the first British settlement began near today's city of Sydney. The settlers were convicts from British jails. The early role of Australia as a kind of outdoor prison had a profound effect on its later character. After serving their terms, the prisoners could buy land and remain. They gave the country a freewheeling, rowdy spirit.

Other settlers followed, although the bleak, dry land on the other side of the world was never popular. "Transporting" prisoners continued until 1868. Because of the sparse rainfall, farming was difficult. The settlers imported sheep from South Africa to begin the colony's major industry. Sheep stations in the outback were huge, because grass was sparse. It took 40 acres to feed a single sheep.

The discovery of gold in the 1850s created Australia's gold rush and tripled its population. It was at this time that Australians began to call each other "digger." The rough-and-ready miners, like the earlier prisoner "transports," had little respect for law and order. It is significant that the most famous Australian of the 19th century was Ned Kelly, a bandit. He became a folk hero, though more like Jesse James than Robin Hood.

Its history shaped Australian society in favor of the ordinary citizen. Trade unions were strong—and still are today. In contrast with the Confucian work ethic that influences many other Pacific Rim nations, Australia's workers have the highest number of paid holidays in the world.

The country's egalitarian character created many benefits. It was the first country in which women gained the right to vote. The secret ballot was developed here—called the "Australian ballot." Today, by law, all Australians must vote, or pay a fine. The strong union movement led to social legislation regulating working conditions and requiring old-age pensions.

A strong economy. In 1901, the six British colonies on Australia united into the Commonwealth of Australia. Australia became an independent nation, although the British monarch, represented by a governor general, remains the official chief of state. The actual head of government is a prime minister who runs the government with a democratically elected parliament.

After World War II, the government surveyed the continent to find hidden resources. The vast interior proved richer than its bleak landscape indicated. Australia now leads the world in the mining of bauxite, from which aluminum is made. Australia is the world's fourth largest miner of both iron ore and gold. Australia's oil fields today produce about 90 percent of its petroleum needs. Australia mines a great deal of coal. The Argyle

These workers use giant trucks to haul coal from mines in Australia—the world's seventh largest coal-producing country.

Mines include one of the largest diamond mines in the world. Most of the world's opals come from Australia, and it also has sapphire fields. Thailand's jewelry makers are the greatest customers for Australia's precious minerals. Land, much of it still unpopulated, is Australia's greatest resource. It produces enough food to give its people a high standard of living. Huge cattle and sheep stations produce hides, wool, and meat to export. Agriculture has thrived, with wheat and other grains, fruits, and vegetables now grown in abundance. Large vineyards have supplied a wine-making industry.

The symbol of the new Australia is Sydney, its largest city. In recent years, the city's waterfront was rebuilt. A grand new opera house stands on the edge of the harbor, with its roof looking like billowing sails. It has attracted musical talent from all over the world.

Sydney's friendly rival is the second-largest city, Melbourne. While Sydney is "flash" and lively, Melbourne is conservative and businesslike. The cities represent two sides of modern Australia. Yet the smaller cities of the northern coast may represent Australia's future. It is there that immigrants from other Asian countries have come to help build new industry.

In the 1970s, Australia was buffeted by lower prices for its minerals and agriculture. The country watched its Asian neighbors, such as Singapore and Japan, become economic giants. It felt left behind and began to create greater industrial development.

High tech has come to Australia in the form of electronics factories and computer plants. Its national educational system has produced a skilled work force. However, higher education is not as strong as in other economies of the Pacific Rim. In Japan, 94 percent of 17-year-olds are still in school—the figure in Australia is 40 percent. All of Australia's 20 universities are run by the government.

Joining Asia. Like the United States, Australia is a land of immigrants. Through the first half of the 20th century, the majority of Australians were descendants of people of Britain and Ireland. The government discouraged other immigrants. There was an unofficial "whites-only" policy.

After World War II, Australia accepted war refugees from eastern and southern Europe. With its wide-open spaces and rich resources, Australia was a promising place to start a new life. The new immigrants enriched the country with their professional and business skills.

In the 1960s and 1970s, with the Asian economy booming, Australia's immigration policy changed. Asians started to arrive in greater numbers. "Boat people" fleeing the wars in Southeast Asia looked for a new homeland. Australia took in a greater percentage of Vietnamese refugees, in proportion to its own population, than any other nation.

The new arrivals proved to be valuable citizens. Their children excelled in school. Carrying the Confucian ethic

Family barbecue, Melbourne, Australia.

with them, many Asians brought a new spirit to Australian business. Finally the Australian government opened its doors to people of many different origins who could contribute to the country. By the 1980s, Australia was a changing nation, though many Asians still faced prejudice from those of European descent.

Australia's foreign policy has also turned toward its neighbors. Realizing that it is closer to Asia than Europe, it has become more involved with Asian affairs. It cooperated with ASEAN in trying to find a settlement of the conflict between Vietnam and Kampuchea. It has increased its trade with other Asian nations. Recently, it has negotiated with Indonesia to solve a crisis in East Timor. Australia has given aid to developing countries in the region.

Despite recent changes, Australia remains a sleeping giant. The countries around it have moved faster. Its extensive social programs have taken money away from needed development. Much of its mineral resources are still undeveloped. Its future success depends on greater participation in the growth of the Pacific Rim.

The Kiwis. Twelve hundred miles east of Australia are the two islands that form the nation of New Zealand. Roughly the size of Colorado, it is a land where sheep outnumber people by 20 to 1. Because the North Island and South Island are long and narrow, everyone lives within 80 miles of the ocean.

The first arrivals to New Zealand were the Maoris,* a Polynesian people. They arrived in great sailing canoes. They gave the land the name *Tiritiri o te Moana**—"gift of the sea."

In 1642, a Dutch ship commanded by Abel Tasman sighted the islands. But when four of Tasman's crew tried to land, the Maoris drove them off. In 1769, Captain James Cook explored the islands. He was fascinated by the strange plants and birds that were unique to the islands. One of them, a bird with a long beak and short legs on a plump round body, was the kiwi. New Zealanders have taken the bird as a national symbol and call themselves "the Kiwis."

In 1840, Britain claimed the islands. By the Treaty of Waitangi, the Maoris agreed to respect the authority of the British crown in exchange for a guarantee of their land rights. Despite conflicts over land between the Maoris and the *pakeha*,* or whites, relations between natives and settlers were better than in Australia. New Zealand gained representative government in 1853 and was granted dominion status in 1907. Today it is an independent member of the Commonwealth.

The country that emerged was in many ways a cozy copy of Britain. Flocks of sheep and herds of cattle graze on its steep, grassy inland hills. Their products became the backbone of the economy. Like the Australians, New Zealanders have a strong egalitarian tradition. Farmers and strong labor unions dominated politics, and women gained the right to vote early in the 20th century.

Though Wellington is the seat of government, the city

of Auckland is the center of the economy. Its two harbors make it the country's busiest port. Auckland is also a lively cultural center and draws ambitious and talented people from both islands—including Maoris, who make up a sizable part of the city's population. Its industries and port have also attracted immigrant workers from Polynesian islands throughout the South Pacific. For this reason, it has been nicknamed the "capital of Polynesia."

Great Britain's farm. New Zealand's economy was so closely tied to Britain that it was often called Great Britain's farm. Even after World War II, Britain was its most important trading partner for 25 years. In 1950, almost 70 percent of New Zealand's exports went to Britain—mostly beef, lamb, dairy products, and wool. This close relationship brought wealth to New Zealand. It enabled the government to build an extensive social welfare system that provided workers with health services and unemployment insurance.

The 1970s was a decade of shocks for New Zealand. It lost the security that its close relationship with Britain had provided. In 1973, Britain entered the European Common Market. This ended the preferential treatment New Zealand's agricultural products had long enjoyed in British trade. The oil crisis hit New Zealand hard because the country lacked its own energy resources.

Prime Minister Robert Muldoon's government reacted by increasing government control of the economy. It increased subsidies to farmers and ranchers, and levied high tariffs on foreign goods to protect New Zealand's industries. Imports dropped drastically. But the resulting cost of such goods as TVs, washing machines, children's clothes, and even shampoo, made them high-priced, luxury items. In response, the government began wage and price controls.

New Zealand's major exports include meat and wool from its many sheep.

The government also encouraged heavy development of alternate energy sources. To pay for these projects while maintaining the high level of social programs, the government borrowed heavily. Its high tariffs isolated the country from the world's markets and weakened the economy. As the debts mounted, the economy was in crisis—10 percent of the GNP went to pay interest on the debt.

Rogernomics. After elections in 1984, David Lange became prime minister. He appointed Roger Douglas as his finance minister. Douglas completely reversed the economic policies of the past. He cut all subsidies for agriculture and other businesses. As a result, the prices of land and livestock plummeted. Government-owned enterprises were sold to private citizens. Import restrictions were lifted, and the financial markets were deregulated. Douglas also cut income taxes—the top rate fell from 66 percent to 33 percent. To make up for that, a flat 10 percent charge was put on all goods and services. New Zealand went from a welfare state to a free-market economy in record time. The sweeping new program was dubbed Rogernomics.

The effects were startling. At first, the stock market soared, the inflation rate came down, and the government had a budget surplus. Exports increased. New Zealand's financiers and business people applauded Rogernomics.

After the first flush of success, however, Rogernomics came under attack. Some had suffered during the changes. Hard-hit farmers claimed that they were carrying the recovery on their backs. The unemployment rate started to climb, reaching over 6 percent in 1988. But among Maoris, it was an appalling 23 percent. The unevenly distributed benefits of Rogernomics threatened the social and racial harmony of the country. Under pressure, Prime Minister Lange fired his finance minister.

Though the pace of change has slowed, New Zealand now has a firm place in the international economy. No longer "Britain's farm," it sends only 9 percent of its exports to Britain. Today, New Zealand's three most important trading partners are Japan, Australia, and the United States.

Prime Minister Lange's foreign policy has created hard feelings with the United States. "No nukes" is the message from New Zealand. Lange has declared New Zealand a nuclear-free country. He denied the use of the nation's port facilities to American navy ships carrying nuclear weapons. In retaliation, the United States suspended its defense agreement with the country.

Around 1900, British author Rudyard Kipling, wrote lyrically about New Zealand: "Last, loneliest, loveliest, exquisite, apart... the Happy Isles!" Much of that description remains true, but it is no longer lonely in its place on the Pacific Rim.

Double-check

Review

1. Define the term *Maori*.
2. What are three major mineral resources of Australia?
3. List three important agricultural products of Australia.
4. Name one of New Zealand's three most important trading partners.
5. What does the phrase "Britain's farm" mean?

Discussion

1. A much lower percentage of 17-year-olds are in school in Australia than in Japan. What effect might this low enrollment play in the future standard of living of Australia? Support your opinion.

2. What are some of the basic natural resources that make it easier for a country to become an economically powerful country? Which of these resources do Australia and New Zealand have? How might this information help you make a prediction about these two countries' economic future?

3. New Zealand fought on the side of the United States in both world wars and signed a military treaty with the United States and Australia in 1951. New Zealand's refusal to allow U.S. ships with nuclear weapons from using its ports caused an end to that alliance in 1986. New Zealand did join with Australia and Britain in an agreement to help defend Singapore and Malaysia if those countries were attacked. Should New Zealand be encouraged to change its policies and allow ships with nuclear weapons within its ports? Why or why not?

Activities

1. Several students might use the *Reader's Guide to Periodical Literature* to locate recent articles about Australia and New Zealand. Using such library resources, students might present a current economic events report on each country to the class.

2. Students might prepare a panel discussion about the role of sports, leisure time activities, and tourism in the Australian economy. Panel members might also recommend whether these industries or heavy industries or high-tech industries should be emphasized in the future.

3. What is it like to live and work in Sidney or Melbourne, Australia, or in Auckland or Wellington, New Zealand? Several students might write to the Australian or New Zealand embassies in Washington, D.C., or to these countries' missions to the United Nations in New York City, in order to obtain this information.

Skills

READING A MAP

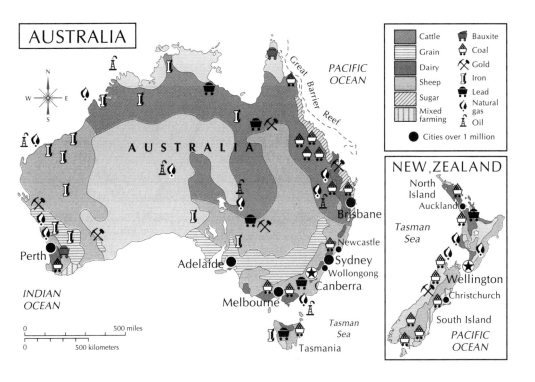

Study the map. Then decide whether each of the following statements is true or false. If the statement is true, write true *next to the statement's number on a separate sheet of paper. If the statement is false, rewrite the statement to make it true.*

1. Most of Australia's and New Zealand's large cities are located near the coasts.

2. Australia's commercial grain crops such as wheat are grown mainly in Australia's north.

3. New Zealand's sheep raising ranches are located only on South Island.

4. Sugarcane is grown along Australia's northeastern coast.

5. Much of Australia's coal is located in eastern Australia.

Chapter 13

Developing Countries of the Pacific Rim

ONE OF AMERICAN AUTO COMPANY chairman Lee Iacocca's greatest fans is Zhang Guozi of the People's Republic of China. Zhang claims that Iacocca's autobiography is the only book he has read in years. A very rich man in the People's Republic today, Zhang is one of the new entrepreneurs who are transforming China's economy. Less than 20 years ago the People's Republic was known for its fanatic devotion to the Marxism of Chairman Mao.

China is one of the nations on the Pacific Rim that is scrambling to catch up with Japan, the Four Tigers, and the Little Tigers. Most of the countries that are working to improve their economies are called LDCs (less developed countries). In addition to China, they include Burma, Vietnam, Kampuchea, and Laos. All have aver-

A communist government poster in Ho Chi Minh City, Vietnam. Despite such posters showing happy farmers and workers, Vietnam's economy remains a very poor one.

age per-person incomes of less than $500 a year. North Korea is considered a middle-income LDC, since its average per-person income is a little higher.

Also on the Pacific Rim, of course, is the Soviet Union, which stretches all the way to Europe. Even though its economy is the world's second largest, it is having some growing pains. Like the LDCs, it may give up some government control of its economy in order to share in the region's rapid development.

Changes in China. Zhang made his fortune in woodworking and furniture and cabinet making. His wood carvings soon picked up export orders through state-controlled Shanghai trading companies. Later he attracted Japanese buyers for what would be his biggest money-maker—wooden Buddhist shrines for homes. Their high quality enabled him to charge $10,000 for each shrine.

Zhang's business grew by leaps and bounds. This earned him criticism as the "tail of capitalism" in the radical days of the mid-1970s. Somehow he managed to survive with the help of friends who were officials in the Communist party.

Today, Zhang has expanded into such other fields as clothes, filmmaking, and real estate. He employs more than 30,000 workers in 30 factories. His life-style flaunts his wealth. He likes nothing so much as slipping a tape cassette into his Japanese-made car and turning the player up full blast as he drives the narrow streets of Yujiang County. Sometimes he has to swerve to avoid water buffalo or donkey carts. "This country is poor, but my world is rich," boasts Mr. Zhang.

Mr. Zhang and the People's Republic are only a part of the new spirit that is sweeping through Asia. Japan and the Four Tigers have demonstrated what is possible in economic development. Their success, as well as that of the Little Tigers, has affected the economies of all of Asia. Every country is in search of new markets and industrial development. Asia, which has 60 percent of the world's population, is increasing in prosperity and pride.

The four modernizations. In the early 1970s, while Confucius was inspiring economic growth in China's neighbors, the ancient philosopher was bitterly denounced in the land of his birth. Communist party members valued "redness," or ideological purity, over expertise and economic success. But after the death of

China's Women's Basketball Team receives pointers from their coach before a game with other Asian teams. In what ways has China's new economic policies affected people's daily lives?

Mao Zedong in 1976, a new era began in the People's Republic. With the rise to power of Deng Xiaoping, the country settled down to try to build a better life for the more than one billion people of China. Reversing Mao's destructive policies, Deng promised to use whatever methods produced results. "Seek truth from facts," was Deng's slogan.

Deng pursued the so-called "four modernizations"—in agriculture, industry, science and technology, and defense. His goal was to quadruple the living standard of the average Chinese by the year 2000. By the year 2049, one hundred years after the founding of the People's Republic, Deng hoped that China's living standard would match that of the developed countries.

To accomplish these goals, Deng decided that farmers and workers had to be given incentives to increase their production. These included higher pay for more productive workers and the development of markets to guide the development of enterprises. These moves, which ran counter to Mao's ideology, would have been impossible earlier.

Deng's government abolished the communes, or large state-regulated farms, that were the centerpiece of Mao's economic policy. Farmers were allowed to make decisions for themselves rather than follow the commands of the state. Communist government control did not disappear—farmers still had to deliver to the state a quota of food. But they could sell any surplus either in a private market or to the state. Rural households were allowed to start their own businesses to supply such needs as agricultural supplies, fixing farm equipment, and consumer goods. This introduced the entrepreneur, or small-business person, to the countryside. Production soared.

In the cities, the People's Republic borrowed an idea from its capitalist rival Taiwan. Special economic zones, like Taiwan's export processing zones, were established in selected areas of the country. In these zones, foreign equipment came into the country duty free. China also encouraged joint ventures with foreign nations to build plants that would help the country develop such sophisticated high-tech businesses as computers.

"I'd heard that it was easy to make money here." The new policies were coupled with a program that beefed up the education system. The government sent students to the finest universities in the West. Throughout the country the curriculum was updated to encourage excellence rather than conformity. Scientific research centers increased, and Chinese scientists were encouraged to participate in international conferences.

The results of these efforts were impressive. After a slow start, the economic zones started to produce high-tech products as well as consumer goods. The lure of a market of more than one billion people drew many foreign investors. As wages gradually rose in the Four Tigers, business leaders looked elsewhere for low-wage workers—China was one of their targets.

The economic growth accelerated. The private-enterprise segment of the economy grew by a staggering rate of 20 percent each year. Changes could be seen on the streets of China's cities. Western fashions and blue jeans replaced the uniforms of the Mao era. In Beijing, the capital, an American fast-food restaurant opened an outlet with seating for 500 and was a smash success. The government even issued to a small number of business people that symbol of capitalism, the credit card—called Great Wall Credit Card. The Great Wall label was also used for China's first computer to enter the international market.

The economic zone areas, particularly around Guangzhou, exploded with people rushing in from impoverished farm areas. "I'd heard that it was easy to make money here," said one new arrival to Guangzhou. For a short period in 1989, as many as 100,000 new people came to the Guangzhou area every day.

The new economic policies brought other serious problems. Inflation rose as did corruption among government officials. Those who did not share in the prosperity resented others who did. The debate grew. Some argued for a return to stronger state control. Others approved of policies that favored more competition and free markets.

In 1989, the government moved to re-establish some governmental direction of the economy. At the same time, the government reassured people that economic reforms would continue. The final result is yet to be seen. In 1989, millions of students and workers gathered in huge protest demonstrations. They demanded more democracy. Chinese leaders ordered the army to crush the demonstrations. Many people were killed.

The LDCs of Southeast Asia. Burma is another of the Pacific Rim's LDCs. Though not actually bordering on the Pacific, it shares the northern part of the Malay Peninsula with Thailand. After Burma gained independence from Britain in 1948, many thought the country had a promising future. It had rich resources and a high literacy rate, but failed to take advantage of them.

Two-thirds of Burmese workers are farmers, and four out of five people live in small rural villages. The main crops are rice, sugarcane, corn, and peanuts. Burma is a main supplier of the narcotic drug opium, which comes from poppy plants. Burma's natural resources include valuable forests from which teak and other wood is cut.

A military strongman, General Ne Win, closed the country to foreigners and persecuted Chinese and Indian minorities. Ne Win's government's heavy-handed control of the economy created stagnation and poverty.

At the end of this street in Ho Chi Minh City, Vietnam, is the historical museum built according to the plans of a pagoda.

Today a new military dictatorship has begun seeking foreign investment and is raising money by selling lumber concessions in Burma's vast forestland. In 1989, the government gave Burma a new name–Myanmar.

In the three countries of Indochina—Vietnam, Kampuchea, and Laos—Communist revolutions succeeded in the 1970s. But the Communist military victors found that it was easier to win a war than to run a country. While the rest of the Pacific Rim region experienced progress and prosperity, Indochina was left behind.

Vietnam rejected both capitalism and its Confucian roots just as capitalism and the Confucian ethic were transforming other countries. Today Vietnam is starting to experiment with less government control over the economy. The government is allowing small businesses to develop, as the Chinese government did earlier.

223

The long period of warfare from the 1940s through the 1970s greatly weakened the Vietnamese economy. In the 1980s the government continued to spend about half its budget on the military. Seven out of every ten workers are farmers, with rice as the main crop. Food processing and textiles are major industries. Heavy industry, such as cement and steel production, is centered in the north.

Vietnam's leaders are particularly concerned about the growing disenchantment of its teenagers and young adults. In this age group, unemployment and crime show sharp rises. There is little interest in sacrificing for the good of communism. Japanese motorcycles and American blue jeans are the highest status symbols for those who can afford them. Ha Quang Du, secretary of the Communist Youth Union, complained that young Vietnamese "want Western-life styles and are always looking for overseas jobs."

Kampuchea (Cambodia) is a war-torn Communist country with an agricultural economy. Continued conflicts between rival Communist factions and with neighboring Vietnam have reduced the people's living standards to among the lowest in the region. Hundreds of thousands of people have fled starvation and political repression in the 1970s and 1980s.

Laos, Southeast Asia's landlocked LDC, has a Communist-run government. Most people are farmers who grow rice or other crops including corn, tobacco, and opium poppies. It remains quite isolated from the rest of the world, but is under the political influence of neighboring Vietnam.

North Korea's industrialization. After the Korean War, mountainous North Korea remained under tight Communist rule. Natural resources such as coal and iron ore were used to develop heavy industry. Hydroelectric projects have long provided important energy sources. North Korea continues to favor reunification with South Korea but does not wish to give up its government-controlled economy for the competitiveness of South Korea's capitalist system.

Perestroika and glasnost. The victory of the free market can be seen most dramatically in recent developments in the Soviet Union. The home of the Russian Revolution has watched its Asian neighbors flourish through their free-market miracle economies. At the same time, the Soviet economy has been stagnating. The domestic economy grew slowly, and shortages of consumer products produced long lines outside stores

The Baikal-Amur Railway. In 1984 workers completed this important railway link between the Soviet Pacific coast and the interior of Soviet Asia.

that had anything to sell. Even worse, the country has had to import food to feed itself, despite the fact that its Ukraine wheat fields were once one of the world's breadbaskets.

When Mikhail Gorbachev came to power in 1985, he promised change. Gorbachev, then the youngest member of the ruling Politburo, risked opposition from the then-entrenched "old guard" Communist leadership. However, he rallied support in his attempts to turn the country around. He called the new approach *perestroika*,* or "restructuring."

Perestroika called for a loosening of government controls and limited markets for free enterprise. Gorbachev reached out for credit from the West, seeking joint enterprises for development. Admitting the failure of Soviet agriculture, he allowed farmers greater freedom in growing and marketing their goods. The program has had mixed success.

Gorbachev also introduced *glasnost*, a new policy of openness to criticism. For the first time, Soviet news programs reported disasters and policy disagreements within the leadership. Small journals printed articles by dissenters whose work had been banned. In 1989, Soviet elections offered a greater choice of candidates than ever before in Soviet history.

The Soviet Union has been developing its vast resources in Asia. Most of this oil, gas, and mineral wealth is in Siberia, but far from the Pacific coast. Some coal, tin, and oil is nearer the coast. The region's leading product is fish. Soviet fishing fleets make huge hauls of herring, cod, and mackerel, which are frozen, canned, and carried by trains to Europe.

In a speech at the major Pacific port of Vladivostok, Gorbachev reminded the world that the Soviet Union is a Pacific power. His peaceful gestures have brought some response. For the first time there has been contact

AVERAGE INCOME PER PERSON IN PACIFIC RIM ECONOMIES

AVERAGE ANNUAL INCOME (in dollars)

| More than $10,000 | $5,001 to $10,000 | $1,001 to $5,000 | $426 to $1,000 | Under $426 |

with the militantly anti-Communist South Korean government, and moves to establish diplomatic relations have been made.

Gorbachev also moved to repair relations with the People's Republic of China. The two nations had been quietly hostile since their troops fought along the border in the early 1960s.

Double-check

Review

1. What is an LDC?
2. What is the major port of the Soviet Union on its Pacific coast?
3. What major food crop is grown in the countries of Southeast Asia?
4. List the "four modernizations" that China hopes to achieve.
5. Define the term *perestroika*.

Discussion

1. Do you think that policies of *glasnost* and *perestroika* will help the Soviet Union develop its economy? Support your opinion.

2. List three reasons why the LDCs of the Pacific Rim have not achieved the economic progress that Japan and the Four Tigers have. Then decide which is the most important reason and explain your reasons for believing so.

3. By comparing resources and industries, formulate several ways North Korea and South Korea might benefit from reunification if political agreement could be reached between the countries.

Activities

1. Several students might select different Pacific Rim developing countries and draw a mural for each one, depicting the major components of its agricultural and industrial economy. The murals might be displayed on the bulletin board.

2. Several students might conduct a panel discussion, with students role-playing representatives of the People's Republic of China, the Soviet Union, and Vietnam at a conference discussing economic development. After doing library research to support their role-playing, each representative should present his or her government's latest economic policies to the rest of the panel and to the class.

3. Several students might conduct a debate to consider whether economic progress depends more on a country's economic system, its people's traditions, or the available natural resources.

Skills

CHINA'S FOREIGN TRADE

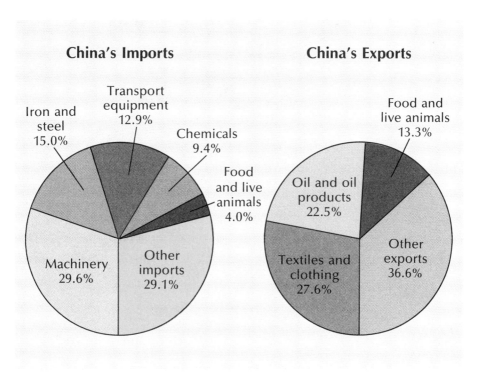

SOURCE: U.S. CIA *Handbook of Economic Statistics, 1988.*

Study the two graphs. Then answer the following questions.

   1. What is China's largest export?

   2. What is China's largest import?

   3. Are China's imports of food and live animals greater or less than its exports of food and live animals?

   4. Is China a major importer or a major exporter of oil and oil products?

   5. Do these graphs show which are China's major trade partners?

Chapter 14

The Eastern Pacific Rim

IN RECENT YEARS, AMERICANS HAVE become used to seeing Asian business executives working in the United States. And Americans are heading the other way—into the Asian Pacific Rim countries. The State Department reported that in 1988 there was a 30 percent growth in Americans living in Asia. Many American firms have regional offices there. But in recent years more firms have opened up and many American firms have gone into partnership with local businesses.

The experience of working in Asia can cause culture shock for an American. Christine Houston, who worked in Japan for seven years, realized that she had to change her way of operating. "When I went to Japan I was a very loud, aggressive 26-year-old New Yorker," she noted. "And I quickly discovered that kind of attitude

An auto plant in California. It is run by an American company and a Japanese one working together.

does not play well in Tokyo." She realized that to succeed she had to become more "low-key" and adapt to the group attitudes so highly prized in Japan.

Christine's experience has been shared by many other Americans. It is not only learning the language that is difficult, but adjusting to different customs. Moreover, the high cost of living in such foreign capitals as Tokyo can be a shock. Many companies must help their employees by paying for their apartments.

Still, the experience can be rewarding. And in the future it will attract more and more Americans, just as many Asian businesspeople are coming to the United States, Canada, and Latin America. The countries on both sides of the Pacific are part of the world's greatest trading region. As economic activity increases, it will change life for people who live in the eastern Pacific Rim as much as it has the Asian side.

From benefactor to partner. No nation has been challenged more by the rise of the Asian Pacific Rim than the United States. Countries that formerly received our economic and military aid now have grown wealthy by competing with, and sometimes destroying, U.S. industries. The United States has seen its role change from benefactor of war-torn Japan to economic partner. The former United States ambassador, Mike Mansfield, said the United States and Japan had "the most important bilateral relationship in the world." Managing that relationship is a challenge for American leaders.

President George Bush acknowledged Japan's new importance by welcoming its prime minister, Noburo Takeshita, as his first foreign guest at the White House. Soon afterward, the President himself paid his respects at the funeral of Emperor Hirohito. Presidents rarely attend the funerals of foreign leaders—this is usually left to the vice president or secretary of state. But the U.S.–

President and Mrs. Bush attend the funeral of Emperor Hirohito of Japan. Among the world leaders shown is Indonesian President Suharto and his wife on the left.

Japan relationship is so important to the United States that Bush represented the nation himself.

Bush's trip to Asia included stopovers in the People's Republic of China and South Korea. Though China is not, like South Korea and Japan, an economic rival, all three countries are important to the United States' national security needs. "We need strong allies," said Anthony Albrecht, a former State Department official. "We need a strong, cooperative relationship."

The United States needs Pacific allies because its chief international rival, the Soviet Union, is also a Pacific power. Since the 1970s, the Soviets have increased the size of their Pacific fleet. According to the Department of Defense, the Soviet Union has four times as many warships in the Pacific as the United States.

The United States and Japan must forge a new cooperative relationship. The United States can no longer dictate policy. "Japan is too big to be under the U.S. umbrella and too vulnerable to be on its own," says

Zbigniew Brzezinski, former American national security advisor. "We need each other," he adds. On the other hand, Japan has to be sensitive to American resentment over its trade dominance. Some officials have suggested that there should be greater coordination of economic policies between the two countries.

Internationally, Japanese financial power has increased its potential for world leadership. But financial power alone does not bring respect. Says Jiro Tokuyama of Mitsui, "We can sell cars and VCRs, but who loves Japan?" And, many believe, Japan is unwilling to assume a leadership role. Despite Japan's growth, the United States still produces about one-fourth of the world's goods and services, compared with Japan's one-tenth. The United States is still looked to as the world's economic leader. But Japan is rapidly catching up.

What happens to the American dream? "At the rate things are going, we are all going to wind up working for the Japanese," says economist Lester Thurow. The huge trade deficit that the United States runs with Japan and other Asian Pacific Rim nations has caused great concern. It accounts for more than half of the total trade deficit of the United States. The deficit comes from the fact that Americans are the world's greatest consumers, and they have acquired a taste for imported goods. In addition, the United States has become the world's greatest debtor nation. These conditions could threaten the American dream itself.

The success of the Asian Pacific Rim countries has been blamed for the loss of manufacturing jobs in the United States, Millions of U.S. workers in auto plants, textile mills, and other industries have lost their jobs because of competition from less expensive or higher quality imported goods. Many of these jobs, such as those in the automobile industry, were highly paid.

A Roll-On/Roll-Off (RO/RO) ship unloading imports at the port of Long Beach, California. The port is the busiest one on the U.S. Pacific coast.

Many of the laid-off workers have not been able to get new jobs that paid as well as their old ones. Most have had to take jobs in the service sector, which often pay much less. Service sector workers include salespeople, clerical workers, repair people, as well as workers in service industries such as banking, insurance, and health care.

Some critics fear that as the manufacturing sector erodes, we will become a nation of hamburger flippers. High-tech industries—in which the United States is still dominant—require a high level of education and do not provide a large number of jobs.

Economic policy in the last decade of the 20th century has become as important as national security. In the halls of Congress, there is agreement that something must be done. In 1988, Congress passed the Omnibus Trade Bill. This legislation required the U.S. government to retaliate against countries that engage in unfair trade practices, such as "dumping" goods—selling products for less than the cost of making them in order to gain a

235

market foothold. The bill targeted countries that kept American businesses from gaining equal access to their markets. The law requires the president to report on trade conditions to Congress on a regular basis.

However, others argue that such laws are too protectionist. If the United States does not allow all foreign goods to enter its borders, American manufacturers and workers might not feel the need to compete to produce high-quality, inexpensive products.

Some have suggested that the United States adopt Japanese methods. They call for more economic planning at the national level and for U.S. companies in the same fields to cooperate. In the past, such cooperation has been forbidden by the U.S. laws against business monopolies. It runs counter to the American idea of competition in the market. Other Japanese methods also do not take into account the American individualism. These include the Japanese workplace model of lifetime employment and company loyalty. Most analysts believe that the United States will have to use its own strengths to solve the problem.

The United States remains open to new immigrants. In recent years, Asians and Hispanics have formed the largest groups. Most arrive with a firm belief in the American dream. In the past two decades, American universities have trained many students from Taiwan, Hong Kong, South Korea, and Japan. These students often remained in the United States to work as engineers and scientists. Now their home countries are trying to lure them back. A recently set up university in South Korea hired 140 Korean faculty members, 120 of them from American universities and labs. However, most immigrants remain in the United States. These immigrants energize the country with their talents. They help ensure that the entrepreneurial spirit is alive and well in the United States.

New challenges. As in Japan, automation and new technology offer gains in productivity and the means for American business to compete more successfully. Many more businesses are using robots for manufacturing and computers for design purposes. They train workers in modern methods to ensure high-quality parts. In the past, it has taken American manufacturers too long to plan and develop products. Tying all these systems together will produce a flexible manufacturing environment capable of producing goods on short notice.

This offers hope for the United States because it will cut the foreign advantage of low-wage workers. These new automated systems will require fewer workers, and costs of manufacturing will go down. Thus the difference between a high-wage and low-wage country will not be so important.

The economies of the west coasts of Canada and the United States are becoming more closely linked with the Asian Pacific Rim. The ports of Vancouver, British Columbia, and Long Beach and Los Angeles, California, are busy with imports and exports. When the economies of East Asia boom, those countries are able to afford to buy American and Canadian goods and services. They are also able to buy raw materials such as timber from Oregon and Washington.

North and south of our border. Japan is the second-largest trading partner of the United States. The first is our northern neighbor, Canada. On January 1, 1989, the United States and Canada put into effect a free-trade agreement. There will now be no tariffs on either country's goods in trade between the two. The agreement was spurred by the keen competition of the Asian Pacific Rim nations and the proposed European free-trade zone scheduled to start in 1992.

Canada, with the world's second-largest land area, has

only 25 million people. But it has virtually 100 percent literacy and rich natural resources, including abundant supplies of nickel, zinc, gold, copper, lead, and silver. Canada also grows wheat and other grains for export.

Traditionally western Canada was less developed, but the surge of economic power to the Pacific Rim has stimulated its development. Today, the west is the fastest-developing region in Canada. Some middle-class residents of Hong Kong, fearing the return of the colony to China, have emigrated to Canada. Most have settled in the city of Vancouver, attracted by its multi-ethnic population. "It's quite acceptable here for different races to work harmoniously together," says new immigrant Danny Gaw. The Hong Kong refugees have rejuvenated the city by using their savings to help fuel industrial development.

Mexico is another Pacific Rim country where trade with the Asian Pacific countries has increased. Almost 90 million people live in Mexico, a land whose resources include some of the world's greatest oil reserves. In recent years, Mexico has developed steel and chemical industries. Export processing zones near the U.S. border produce electronic products. Mexico is the third-largest trading partner of the United States. Mexico's fears of U.S. dominance have hindered U.S. investment.

Mexico's government regulation and control hamper entrepreneurship. The huge government bureaucracy is costly and inefficient. When President Carlos Salinas de Gortari was inaugurated in 1988, he pledged to cut back the government's role in the economy and let free-market forces work. The new president has often mentioned his respect for the Japanese work ethic and hopes to copy the success of the Four Tigers. He believes Mexico can take heart from their progress. After all, twenty years ago no one believed that South Korea would be an economic power.

The storm cloud over Mexico's development, and that of many South American countries as well, is the problem of enormous debt. Countries such as Mexico, Brazil, Argentina, and Venezuela borrowed heavily, mainly from U.S. banks, to develop their resources and industries. The money was borrowed when economic growth was strong and Latin Americans expected that it would continue. The natural resources of timber, minerals, oil, and agricultural wealth promised growing prosperity.

But falling prices for South American commodities made repaying the debt difficult. New loans were sought to pay back older ones. To get the new loans, Latin American governments have had to pledge austerity programs. Social programs and food subsidies were cut, leading to discontent among their people. In 1989, rioters demonstrated for days to protest Venezuela's domestic cutbacks and rising food prices.

Worse yet, the lack of funds and pressure to pay back the loans mean that long-term development plans no longer can be carried out. As a result of the debt crisis, investors from the United States and Europe have become wary of pumping more money into Latin America. This only makes the situation worse.

As a result, South American countries are looking westward to newly rich Japan. Brazil has been the most successful Latin country in attracting development and

Canada's major Pacific port is Vancouver. Through it, Canada exports coal, grain, and lumber to Japan, grain and potash to China, and coal, potash, and sulpher to South Korea.

An iron refinery in Brazil. Although Brazil has no Pacific shoreline, the third largest customer of its exports is Japan. Brazil has even asked Japanese experts about the possibility of building a highway or a canal from the Amazon River through the Andes Mountains to the Pacific coast.

investment money. Brazil has a large Japanese immigrant population, and there are strong ties betrween the two countries. The Latin debt crisis is so big that many believe Japan's help is essential in solving it. This may be the international issue in which Japan will emerge as an international leader.

Latin America is also interested in Japan's new role as a provider of foreign aid. With an economy still less than one-half that of the United States, Japan now gives the most help to countries in need of economic assistance.

Planning for the 21st century. American businesses are already planning for the 21st century. Today's business executives agree that their replacements will have to be different kinds of people.

Those at the top of today's large corporations usually

came from the finance and marketing divisions of the company. The "fast track" to the top was within corporate headquarters. Many corporate heads had little experience overseas. That will change as businesses must adopt a new, international outlook.

More and more, businesses will be multinational. Manufacturing plants will be built in countries where wages are low. Financing will come from Japan and other wealthy Asian and Western nations. Companies will market their products not only in the United States, but in Europe and Asia as well.

In a recent survey, more than 70 percent of American business leaders said that their worldwide employment will increase by the year 2000. More than half said fewer of their workers would be Americans. Even the heads of tomorrow's multinational companies will be people born outside the United States. American companies are training and promoting talented people from many countries.

To succeed in tomorrow's world, America's business executives will have to deal with foreign governments and economic leaders. They will need to know about the history and customs of many countries. Fluency in foreign languages will be useful, if not absolutely necessary. And American workers in all kinds of occupations will find it valuable to understand the interrelationships among the many economies of the Pacific Rim. Douglas Danforth, the former chairman of Westinghouse, advises future business leaders, "Travel overseas."

The 19th-century American Horace Greeley foreshadowed those words when he said: "Go west, young man, go west." Today, women are among those being urged to go west, and the west extends all the way across the Pacific Ocean. Moreover, in a very real sense, those in Asia are looking eastward to North, Central, and South America, and particularly the United States, for education, ideas, technology, investments, and markets.

Double-check

Review

1. Which Pacific Rim country now provides more foreign aid to other countries than any other nation?

2. List, in order, the three leading trading partners of the United States.

3. What is meant by the term *multinational* when it is used to refer to a company?

4. Name two important port cities on the west coast of North America.

5. What energy resource in Mexico might Asian Pacific Rim countries find valuable to their economies?

Discussion

1. Should the United States try to restrict the importing of high-tech products, textiles, and autos from Asian Pacific Rim countries? Why or why not?

2. Both the United States and the Soviet Union have huge naval bases in other Pacific Rim countries. (U.S. bases in the Philippines; Soviet bases in Vietnam). Should these military installations be continued? Support your opinion.

3. What role, if any, should Japan and the Four Tigers play in the Latin American debt crisis? Why?

Activities

1. Several students might role-play advisers to major United States and Canadian corporations. They could favor or oppose the setting up of Japanese workplace models in the United States and Canada, including lifetime employment. Several other students might take the roles of business leaders and question the advisers on their policy suggestions.

2. Several students might conduct a debate on the following topic: Should American companies be free to buy up Japanese and Four Tiger businesses and land? Should Japanese and Four Tiger companies be free to buy up American businesses and land?

3. The class might decide on the three most important ideas they have gained from learning about the Pacific Rim. Then students might select from these three themes and draw a poster illustrating the idea they have selected. The posters could be put on display.

Skills

MAJOR U.S.–PACIFIC RIM TRADE PARTNERS

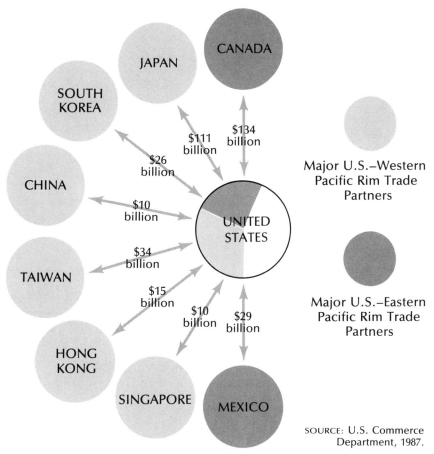

SOURCE: U.S. Commerce Department, 1987.

Study the diagram. Then answer the following questions.

1. Which western Pacific Rim country has the most trade with the United States?

2. Which eastern Pacific Rim country has the most trade with the United States?

3. Which of the Four Tigers has the most trade with the United States?

4. Which western Pacific Rim economies have more trade with the United States than Mexico does?

5. What U.S. government department could provide you with the most recent figures for Pacific Rim trade with the United States?

243

EPILOGUE
THE PACIFIC CENTURY

THIS BOOK HAS OFFERED A BRIEF INTRODUCTION to the Pacific Rim. As we enter the Pacific Century, perhaps it will help you understand the economies and peoples that are shaping it.

Today, you may encounter the influence of the Pacific Rim when you, your friends, or members of your family use a household appliance, ride in a car, or eat a meal.

Hong Kong today.

Many of the jobs in your community are dependent—directly or indirectly—on these economies.

Japan, the Four Tigers, and the Little Tigers offer a challenge to the rest of the world. To developing countries, they offer hope that a similar miracle may lie in store for them. To developed nations, they offer a challenge to established ways of doing business. What can they learn to spur further growth?

Although several of them have experienced phenomenal growth, the countries of the Pacific Rim face many new challenges. As their economies continue to grow, how will they compete and cooperate? To what extent will their governments be affected by this growth? How will the benefits of industrialization be distributed among the region's people?

In this century, several Pacific powers enter the world's economy on equal footing with their former colonial rulers. How will these shifts in power affect these nations? How will Japan and the United States address their trade disputes? How will the Pacific Rim interact with the industrial might of the European Community?

It has been roughly five hundred years since Columbus left Spain in search of Asia. That journey has finally been completed.

Pronunciation Guide

Asian words and names that are familiar to Americans and are found in *Webster's Ninth New Collegiate Dictionary* are accented as the dictionary indicates. Unfamiliar words and names that do *not* appear in the dictionary in some cases are left unaccented, since the subtleties in accenting some Asian languages concern only those who undertake a serious study of the language.

The following system translates each syllable into the nearest common English equivalent. Syllables set in capitals are accented. Principal sound equivalents are:

a (as in cat)
ah (as in odd)
ay (as in ale)
ch (as in chair)
e (as in silent)
ee (as in eat)
eh (as in end)
g (as in go)
i (as in charity)
igh (as in ice)
ih (as in ill)
j (as in John)
k (as in keep)
o (as in connect)

oh (as in old)
oo (as in too)
oo (as in foot)
ow (as in out)
s (as in sit)
t (as in tin)
th (as in them)
u (as in cube)
u (as in circus)
uh (as in unaccented a as in sofa)
ur (as in urn)
y (as in yet)
z (as in zebra)
zh (as in vision)

amakuduri—ah-mah-koo-dah-ree
Amaterasu—ah-mah-teh-rah-soo
archipelago—ar-kuh-PEL-uh-goh
Bao-Dao—BAH-oh DAH-oh

bauxite—BAWK-sait
bumiputera—boo-mee-poo-teh-rah
bushido—boo-SHEE-doh
Chao Phraya—chah-oh PRAH-yah
conquistadores—kon-KEES-tuh-dor-eez
dai josai—dai JOH-sah-ee
daulat—da-oo-laht
Edo—EH-doh
ema—eh-mah
entrepôt—ahn-truh-POH
gin—gin
Ginza—GIN-zah
glaciers—GLAY-shurz
Hangul—HAN-gl
Hokkaido—HOK-KAI-doh
Honshu—HON-shoo
Ijime—ee-jee-meh
jishu kanri—jee-SHOO KAHN-ree
juku—joo-koo
kami—kah-mee
kampongs—kam-pongz
Kaohsiung—KOW-shee-OONG
Kuroshio—koo-roh-shee-oh
kyoiku mamas—KYOH-ee-koo MAH-mahz
Kyoto—KYOH-toh
Kyushu—kyoo-shoo
Malaysia—mah-LAY-zhah
Maoris—MAH-oh-reez
Meiji—may-ee-jee
mobo—moh-boh
moga—moh-gah
monsoon—mahn-SOON
muezzin—myoo-EEZ-uhn
Nagasaki—nah-gah-sah-kee
nemawashi—neh-mah-wah-shee
Osaka—OH-sah-kah
Oyashio—oh-ya-shee-oh
pakeha—pah-keh-ha
perestroika—peh-res-TROI-kah
samurai—sah-moo-rah-ee
Seoul—suh-ool
sha-in—shah-IN

Shikoku—shi-KOH-koo
Shinto—SHIN-toh
shogun—SHOH-goon
Shr-Jin Wei—shoor-jin WEH-ee
Taipei—tai-PAY
Thailand—TAI-land
Tiritiri o te Moana—tih-ree-tih-ree oh teh moh-ah-nah
Tokugawa Ieyasu—toh-koo-gah-wah ee-yeh-yah-soo
Tokyo—TOH-kyoh
Tsun-Mei Chung—Tsoon-MEH-ee CHUNG
urbanization—uhr-buh-nih-ZAY-SHUN
wayang—wah-yang
Yamato damashii—yah-MAH-toh dah-MAH-shee-ee
Yi Song-gye—EE sung-geh
zaibatsu—zai-bah-tsoo

Index

Aborigines, 204
agriculture, 24–27, 147, 195, 207, 210, 222, 225, 226
Akihito, emperor of Japan, 35, 83
Akio Morita, 111
American dream, 234–236
animism, 40–41, 43, 45, 70, 79
Aquino, Corazon, 197–98
archipelagos, 19, 191, 192
Argentina, 239
ASEAN (Association of South East Asian Nations), 62, 188, 209
Asian names, 183
Auckland, New Zealand, 211
Australia, 14, 19, 22–23, 27, 28, 43, 202–215
   climate, 22-23, 204
   immigration policy, 208–209
automation, 237

Bangkok, Thailand, 22, 29, 194, 195
baseball, 136, 158–159
Beijing, PRC, 221
Borneo, 188, 192
Brazil, 239–240
Brunei, 62, 188, 190
Brzezinski, Zbigniew, 234
Buddhism, 41–43, 44, 45, 69–70, 81, 82, 125, 189, 192
Burma (Myanmar), 27, 57, 195, 217, 222–223
Bush, George, 137, 232–233
business leaders, management, 104, 106–107, 240–241

California, 72, 237
Canada, 237–238
capitalism, 59, 61, 63, 111, 223
   Hong Kong, 164, 167, 171
   PRC, 218, 220–221
Caroline Islands, 76

Chiang Ching-kuo, 156,
Chiang Kai-shek, 144–147, 156
China, 57, 90, 91, 125
   civil war, 59–60, 144
   see also People's Republic of China (PRC)
Chinese people abroad, 34, 188–189, 222
Christianity, 43, 45, 70, 71, 81
Chung Ju-Yung, 137
cities, 27–29, 46, 75–76, 86–87
cold war, 59–63
colonies, colonialism, 17–19, 54–57, 58, 61, 92, 146, 176, 188, 190, 222
   Japan, 76, 146-47
   and natural resources, 29
communism, 59, 60, 61–62, 63, 164, 223–224
competition, 14–15, 63, 117, 232, 234-235, 236, 237
Confucianism (Confucian ethic), 34–40, 46, 125, 170, 208–209, 223
   Japan, 34, 39, 69, 81, 82
   Taiwan, 153–157
Confucius, 34, 35–37, 63, 219
Cook, James, 204–205, 210
cooperation, 14–15, 62, 188
culture(s), 46, 49, 125, 192
   see also, individual countries
culture groups, 40

Danforth Douglas, 241
Deming, W. Edwards, 112–113
democracy, 88, 136–137
Deng Xiaoping, 171, 220, 222
developing countries, 217–229
Douglas, Roger, 212–213
Dutch East Indies (Indonesia): *see* Indonesia

250

Eastern Pacific Rim, 231–243
economic development, 64, 218–219
　Australia, 206–208
　Canada, 238
　Hong Kong, 165–169
　Indonesia, 192–193
　Japan, 13, 73–77, 85–101, 107–110
　New Zealand, 212–213
　Philippines, 196–197
　PRC, 221–222
　Singapore, 175, 182
　South Korea, 127–131
　Soviet Union, 225–227
　Taiwan, 147, 148, 149–153
　Thailand, 194–195
economic planning, 110–11, 240–241
economic success
　Confucianism in, 34–40
　effects on U.S., 234–236
　reason for Japanese, 107–110
economic zones, 220–222
economies, Asian, 9–10, 13–14, 55
economy, international, 10, 15, 213, 241
Edo, Japan, 70, 71, 73
education, 48, 235
　Australia, 203, 208
　Hong Kong, 170
　Japan, 74, 113–116
　New Zealand, 204
　PRC, 221
　respect for, in Confucian ethic, 36, 38–40
　Singapore, 181
　South Korea, 134–135, 138–139
　Taiwan, 154
energy sources, 29, 211, 212, 225
Europe, 10
European Common Market, 211
export economies, 14, 128, 166, 178
　Japan, 75, 89, 94–95
　New Zealand, 211, 213
　Taiwan, 147, 148, 150
export processing zones (EPZ), 148, 150, 221, 238

family-owned business, 128–131, 153–154
family relations (Confucian ethic), 37, 40, 69
farmers, farming, 188, 193, 195, 222, 224
　see also agriculture
Federation of Malaysia, 176, 188
foreign investment, 167–168, 192–193, 195, 197, 223
four modernizations (PRC), 219–221
Four Tigers, 13, 117, 129, 182, 188, 217, 218–219
　Confucian ethic, 34
　Little Tigers and, 195
　model for success, 238
　natural resources, 29
　people of, 34
　wages, 221
France, 54, 61
free market, 15, 181, 225
free trade ports, 164–165, 190
free trade zones, 237

Germany, 76, 90
glasnost, 225–227
Golden Triangle, 195
Gorbachev, Mikhail, 226–227
government role in economy, 107–109, 211–212, 222, 223, 225, 238
Great Britain, 17–19, 54
　and Australia, 204, 205
　colonies, 19, 55–57, 176, 188, 190, 222
　Hong Kong and, 164–165, 170–171
　and New Zealand, 210–211
Greater East Asia Co-Prosperity

251

Sphere, 58, 90, 92
Guangzhou, PRC, 221–222

health care, 48
high technology, 96, 117, 128, 134–135, 150, 208, 221, 235, 237
Hinduism, 43, 189, 192
Hirohito, emperor of Japan, 77, 82, 92, 93, 113
  death of, 83, 116, 232
Hiroshima, Japan, 92
history, 53–65, 70–77, 204–206
Hokkaido, 67
Holland, 54
Hong Kong, 9–10, 13, 19, 27, 129, 163–171, 195, 236
  competition with Japan, 117
  as free port, 164–165, 166, 171
  people of, 34
  refugees in Canada, 238
  social change, 46
  standard of living, 48
Honshu, 67

India, 41, 43, 57
Indochina, 90, 91
Indonesia, 13–14, 55, 91, 152, 188, 190–193, 209, 223–224
  agriculture, 26, 27
  in ASEAN, 62
  climate, 22
  culture, 46
  land, 19, 20
  natural resources, 28
  religion, 43
  in World War II, 57
industrial development, 29, 219, 238
  Australia, 208
  Hong Kong, 166–167
  Japan, 75, 76, 94
  Malaysia, 189–190
  North Korea, 224–225
  South Korea, 128–131, 134
  Taiwan, 150
  Vietnam, 224

infant mortality rate, 48
Islam, 43, 45, 192, 198

jaebul, 127–131, 132, 133
Japan, 9–10, 77, 169, 182, 188, 190, 208, 213
  agriculture, 26
  annexed Korea, 57, 76, 125
  climate, 22
  colonialism, 76, 146–147
  emperor, 71, 73–74, 82–83
  governance, 70–71, 73, 74, 82, 83, 88, 90, 93
  history, 57–58, 70–77
  identity crisis, 118–119
  industrial strategy, 95–96, 110–111
  land, 19, 20, 21
  life in 103–121
  militarism, 82, 89–91
  (MITI), 110–111
  as model, 217, 218–219, 236
  people of, 34
  postwar reconstruction, 92–99
  relation with U.S., 14, 71–73, 94, 95, 116–117, 232–234
  religions, 43, 45, 69, 70, 78–81
  as source of aid, 239–240
  South Korea and, 129, 130–131
  standard of living, 48
  trade with West, 54–55, 73
  tradition of borrowing, 68–70, 73, 75
  world role of, 116–117, 234
  and World War II, 13, 61, 125
Java, 191, 192
Joint ventures, 127–131, 190, 196–197, 221

Kalimantan, 192
Kampuchea (Cambodia), 61–62, 209, 217, 223, 224
Kim Il Sung, 126
Konosuke Matsushita, 107–108
Korea, 19, 34, 60, 125–126
  annexed by Japan, 57, 76, 125
Korean War, 60, 125–127, 133,

136, 165, 193–194, 224
Kowloon Peninsula, 165, 167, 170
Kuala Lumpur (KL), Malaysia, 189, 190
Kyoto, Japan, 69, 71, 73
Kyushu, 67, 75

labor costs, 13, 117, 128, 177
 see also wages
land, 19–21, 67–68, 117, 209
land reform, 93, 147
Lange, David, 212, 213
languages, 40, 179, 188, 192, 232, 241
Laos, 27, 62, 195, 217, 223, 224
Latin America, 239–40
LDCs (less developed countries), 217–218, 222–224
Lee Kuan Yew, 33, 176, 177, 178, 181, 182, 183
Lee Teng-hui, 156
life expectancy, 48, 117
literacy rate(s), 48, 154, 222, 238
Little Tigers, 13–14, 29, 187–201, 217, 219
Long Beach, California, 237

MacArthur, Douglas, 61, 93, 126
Magellan, Ferdinand, 53–54
Malay Peninsula, 175, 188, 222
Malaysia, 13–14, 27, 28, 55, 57, 62, 187–190
 land, 19, 20
 religions, 43
Manchuria, 57, 76, 90
Manila, Philippines, 196
manufacturing sector (U.S.), 234–235, 237
Mao Zedong, 144, 146, 171, 217, 220
Maoris, 210, 211, 213
Marcos, Ferdinand, 196, 197
Marianas, 76
Marshall Islands, 76
Melbourne, Australia, 207–208
Mexico, 238–239
middle class, 117–118, 132, 169

modernization 74, 85–88, 129
Muldoon, Robert, 211
multinational businesses, 177, 241
Myanmar. See Burma.

Nagasaki, Japan, 92
nationalism, 58–59, 89, 90, 192
natural resources, 28, 55, 57, 75, 182, 222, 238
 acquired by conquest, 76, 90
 Australia, 29, 206–207, 209
 Little Tigers, 188, 190, 191, 195
 North Korea, 224–225
 South America, 239
 South Korea, 127–128
 Soviet Union, 226
 U.S., 110
Ne Win, General, 222
New Territories, 165, 167, 170
New Zealand, 14, 19, 20, 23, 27, 43, 204, 210–215
newly industrialized economies (NIEs), 13, 129, 194–195
North Korea, 26, 60, 126, 136, 137, 218, 224–225
North Vietnam, 61–62

ocean currents, 22–23
oil, 29, 90–91, 130, 152, 177, 206
oil crisis, 94–95, 128, 211
opium, 27, 195, 222, 224
Oregon, 237
Osaka, Japan, 71, 86

Pacific region, 9–10
Pacific Rim countries, 10–15, 29, 54
 Confucian ethic in, 34–40
 developing countries of, 217–229
 peoples of, 33–50
Panama Canal, 54
Park Chung Hee, 127–128, 129
PRC: see People's Republic of China
Pearl Harbor, 57, 91–92
Penang (island), 189–190

253

People's Republic of China,
    (PRC), 19, 61, 136, 155, 217–222, 233
    and Korean War, 126
    Soviet Union and, 227
    and status of Hong Kong, 164, 170–171
    and Taiwan, 156–157
    trade embargo on, 165
    in United Nations, 148
per capita income, 48, 190, 191, 194–195, 218
perestroika, 225–227
Perry, Matthew C., 54, 71–73
personal savings rate, 109–110, 155
Philippines, 13–14, 57, 62, 188, 196–197
    history, 54, 55, 71
    land, 19, 20
    religion, 43
population, 46, 55–57, 165
    Japan, 57, 67, 86
Portugal, 54, 70
product quality, 10, 99, 111–113
prosperity, 132, 144, 168–169, 182, 219
protectionism, 107–109, 236

Raffles, Sir Thomas Stamford, 176
religions, 40–45, 70, 81–83, 189, 192
Rhee, Syngman, 60, 126–127
rice farming, 21, 24–26, 83, 193, 224
robots, 237
Rogernomics, 212–213
Roh Tae Woo, 136
Russia, 125, 225
    see also Soviet Union
Russo-Japanese War, 57, 76

Sabah, 188
Sarawak, 188, 190
Saudi Arabia, 152
Seoul, S. Korea, 27, 46, 123, 126, 127, 131–133

service sector, 117, 235
Shikoku, 67
Shinto, 45, 70, 78–81
Singapore, 9–10, 13, 21, 129, 169, 175–185, 188, 208
    in ASEAN, 62
    climate, 22
    competition with Japan, 117
    land, 19
    people of, 33, 34
    religion, 45
    social change, 46
    standard of living, 48
    urbanization, 27, 28, 29
social change, 46
    Japan, 73–74, 85–88, 93
South America, 238, 239–240
South Korea, 9–10, 13, 14, 60, 123–141, 225, 227, 233, 236, 238
    agriculture, 26
    climate, 22
    Confucian ethic, 34
    competition with Japan, 117
    land, 19
    literacy rate, 48
    natural resources, 29
    religion, 43
    reunification issue, 136–137
South Vietnam, 61–62
Southeast Asia, 19, 176, 222–224
Soviet Union, 126, 136, 218, 225–227, 233
    and U.S., 59–63
Spain, 53–54, 70, 71
sports, 136, 154, 158–159, 192
standard of living, 46–49, 168–169, 178, 207, 220, 224
Suharto, 192, 193
Sumatra, 191
Sydney, Australia, 27, 205, 207

Taipei, Taiwan, 143, 144–146
Taiwan, 9–10, 13, 14, 117, 129, 143–161, 220–221, 236
    Chinese government at, 60
    conquered by Japan, 57, 76

land, 19
people of, 34
political fate of, 156-157
sports in, 154, 158–159
tariffs, 15, 89, 107–109, 211, 212
Tasman, Abel, 210
technology rights, 109
see also high technology
Thailand, 13–14, 46, 54, 62, 188, 194–195, 222
agriculture, 26
land, 19, 20, 21
natural resources, 28
urbanization, 27
Tokugawa family, 70–71
Tokyo, Japan, 13, 17, 73, 86, 92, 94
Stock Exchange, 99
tourism, 168 178, 197
tradition, 13, 135, 189

unions, 105, 134, 136, 206, 210
United Nations, 60, 126, 146, 149
United States, 11, 48, 55, 117, 146, 153, 196, 231–237
aid to Pacific rim countries, 61, 63, 93, 107, 127, 147, 196, 232
competition with Soviet Union, 59–63
debtor nation, 131, 234
economic leadership of, 234
economic planning, 110–111, 240–241
and Korean War, 126, 127
National Security Council, 111
navy, 77, 91–92
relation with Japan, 71–73, 94, 95, 116–117, 232–234
relations with Pacific Rim countries, 14–15, 130–131, 146, 147, 148, 213, 232–234
trade deficit, 15, 95, 131, 153, 234
trade with Pacific Rim countries, 54, 55, 213

and Vietnam War, 61–62
in World War II, 90–92
urbanization, 27–29, 189

Vancouver, British Columbia, 237, 238
Venezuela, 239
Vietnam, 26, 209, 217, 223–224
Vietnam War, 61–62, 224

wages, 13, 46–48, 75, 237, 241
Hong Kong, 166
Japan, 75, 86, 94
PRC, 221
Singapore, 177, 178
South Korea, 128, 129, 134
Taiwan, 148, 150
of women, 106
Washington (State), 237
Wellington, New Zealand, 210
West Irian, 192
Western influences, 86–88, 115–116, 224
women
right to vote, 206, 210
women in work force 105–106, 148, 169
work ethic, 34, 132–133, 154, 169–170, 182, 238
Japan, 110, 118, 119
work force, educated, 110, 154, 208
World War I, 76
World War II, 49, 57–58, 61, 94, 144
Japan, 13, 82, 90–92, 116, 125
period following, 58–63

Yi Song-gye, 123, 132
young people, 46, 87–88, 118–119, 139, 154, 155–156, 170, 224

zaibatsu, 74–75, 85–86, 88, 89, 93, 125, 128
Zhang Guozi, 217, 218

PHOTO AND ART CREDITS: • 8, © Robin Laurance/Photo Researchers, Inc. • 11, © Allen Green/Photo Researchers, Inc. • 12, © 1981 Bernard Pierre Wolff/Photo Researchers, Inc. • 15, Overseas Information Office of the ROK Government • 16, David Brauchli/Reuters/Bettmann Newsphotos • 20, © 1977 Jan Lukas/Photo Researchers, Inc. • 21, © 1984 Robert Alexander/Photo Researchers, Inc. • 23, Arthur Tsang/Reuters/Bettmann Newsphotos • 24, © Frank Grant/International Stock Photo • 26, © Paolo Koch/Rapho/Photo Researchers, Inc. • 32, © 1987 Joseph Rupp/Black Star • 35, The Bettmann Archive • 38, © 1988 Anthony Suau/Black Star • 39, Stuart Franklin/Sygma • 41, Sadayuki Mikami/AP/Wide World • 43, Thailand Government Tourism Office • 45, © 1984 Michele and Tom Grimm/International Stock Photo • 49, Fr. Tom Donaher, Catholic Foreign Mission Society of America, Inc. • 52, George Holton/Photo Researchers, Inc. • 55, The Bettmann Archive • 58, UPI/Bettmann Newsphotos • 59, Philip Little/International Stock Photo • 60, Stanley Tretick/UPI/Bettmann Newsphotos • 62, UPI/Bettmann Newsphotos • 66, ©1981 Bernard Pierre Wolff/Photo Researchers, Inc. • 68, ©1986 Bill Bachman/Photo Researchers, Inc. • 69, The Bettmann Archive • 74, Orion Press/Scala/Art Resource, Collection of the Ministry of Foreign Affairs • 77, The Bettmann Archive • 78, Horace Bristol, Jr./Photo Researchers, Inc. • 84, © Robin Laurance/Photo Researchers, Inc. • 87, Brown Brothers • 88–89, The Bettmann Archive • 91, The Bettman Archive • 95, Greg Davis/Sygma • 97, ©1986 Robert A. Isaacs/Photo Researchers, Inc. • 102, ©1986 Robert A. Isaacs/Photo Researchers, Inc. • 104, © Paolo Koch/Photo Researchers, Inc. • 105, ©1986 Robert A. Isaacs/Photo Researchers, Inc. Courtesy of Monolithic Memories • 106, ©1986 Robert A. Isaacs/Photo Researchers, Inc. Courtesy of Monolithic Memories • 112, J.P. Laffont/Sygma • 115, J.P. Laffont/Sygma • 119, ©1986 Bill Bachman/Photo Researchers, Inc. • 122, Bei Yeon-hong/AP/Wide World Photos • 127, Shostal • 129, AP/Wide World Photos • 130, Masaharu Hatano/Reuters/Bettmann Newsphotos • 131, ©1987 Anthony Suau/Black Star • 133, Kim Chon-kil/AP/Wide World • 135, ©1988 Comstock/J. Wishnetsky • 137, Yun Jai-hyong/AP/Wide World Photos • 138, AP/Wide World Photos • 142, Pan Yueh-kang/AP/Wide World Photos • 144, Yae Loong Motor Company • 145, UPI/Bettmann Newsphotos • 146, AP/Wide World Photos • 149, ©1986 Joseph Rupp/Black Star • 151, Reuters/Bettmann Newsphotos • 152, Yang Chi-hsien/AP/Wide World Photos • 155, ©1972 George Holton/Photo Researchers, Inc. • 158, Paul Vathis/AP/Wide World Photos • 162, © Robin Laurance/Photo Researchers, Inc. • 166, © Porterfield-Chickering/Photo Researchers, Inc. • 167, Raymond Darolle/Sygma • 169 Stuart Franklin/Sygma • 170, Hong Kong Government Public Information Office • 174, J. David Day/Shostal Associates • 179, © Information Division, Ministry of Communications and Information, Singapore • 180, ©1983 Abraham Menashe/Photo Researchers, Inc. • 182, Dominic Wong/Reuters/Bettmann Newsphotos • 186, Arthur Tsang/Reuters/Bettmann Newsphotos • 189, Bernard Pierre Wolff/Photo Researchers, Inc. • 191 ©1983 Stephanie Dinkins/Photo Researchers, Inc. • 194, ©1988 Mike Yamashita • 197, Claro Cortes IV/Reuters/Bettmann Newsphotos • 202, © Ulrike Welsch/Photo Researchers, Inc. • 205, ©1988 Bill Bachman/Photo Researchers, Inc. • 207, Jean Guichard /Sygma • 209, Australian Overseas Information Service • 212, Compliments of the Consulate General of New Zealand • 216, Mike Theiler/Reuters/Bettmann Newsphotos • 219, Liu Xinning (Xinhua News Agency)/AP/Wide World • 223, Christine Spengler/Sygma • 225, L. Sverdlov/Tass from Sovfoto • 230, UPI/Bettmann Newsphotos • 233, Sadayuki Mikami/AP/Wide World Photos • 235, Port of Long Beach • 239, Vancouver Port Corporation • 240, ©1988 Ulrike Welsch/Photo Researchers, Inc. • 244, ©1987 Comstock • 245, © Marcel Minnée/International Stock Photo. Cover: ©1986 Nik Wheeler/Blackstar. Maps: David Lindroth. Art: Harry Chester Associates.